Exploring
Statistics

STATISTICS: Textbooks and Monographs

A Series Edited by

D. B. Owen, Coordinating Editor
Department of Statistical Science
Southern Methodist University
Dallas, Texas

R. G. Cornell, Associate Editor
for Biostatistics
University of Michigan

W. J. Kennedy, Associate Editor
for Statistical Computing
Iowa State University

A. M. Kshirsagar, Associate Editor
for Multivariate Analysis and
Experimental Design
University of Michigan

E. G. Schilling, Associate Editor
for Statistical Quality Control
Rochester Institute of Technology

ADDITIONAL VOLUMES IN PREPARATION

Exploring Statistics

Damaraju Raghavarao
Temple University
Philadelphia, Pennsylvania

Marcel Dekker, Inc.   New York • Basel

Library of Congress Cataloging-in-Publication Data

Raghavarao, Damaraju.
  Exploring statistics
  p. cm.—(Statistics, textbooks and monographs; v. 92)
  Includes index.
  ISBN 0-8247-7952-5
  1. Statistics. I. Title. II. Series.
QA276.12.R33 1988   519.5—dc19   87-36893

MARCEL DEKKER, INC.
270 Madison Avenue, New York, New York 10016

Current printing (last digit):
10 9 8 7 6 5 4 3 2 1

PRINTED IN THE UNITED STATES OF AMERICA

to
Lord Venkateswara Swamy

Preface

This book is intended to serve two purposes: to enable professionals such as medical doctors, engineers, business executives, laboratory technicians, school teachers, and others to understand the basics of statistical thought through self-study and, for classroom use in teaching a one-semester first course in statistics for non-calculus students. It gives a concise and quick overview of the commonly used statistical methodology. The main text is free from messy formulae and introduces the reader to the concepts through examples involving real or artificial data. The selection of examples is diversified so that the reader can see the breadth of statistical applications. Every chapter after the first has an appendix giving the necessary mathematical details related to that chapter.

Self-studying readers can skip the appendices, read the main text and familiarize themselves with some of the available statistical tools used in analyzing data. They will pick up the necessary terminology and ideas to have a meaningful dialog with the statistician so that they can form a team in solving the problem of interest. This is a reference book for such an audience.

This book can be conveniently and effectively used as a text for a first level statistics course. The instructor can ask the students to read the main text at home, discuss the highlights of the text, and give an excellent classroom review of the appendices. Exercises at the end of Chapters 2 to 7, with answers for students' practice, are kept to a minimum, as it is my belief that students of such courses do not have enough time or aptitude to solve many problems. At the end of the book, 60 multiple choice review exercises without answers are given, and some or all these problems can be used by the instructor as homework assignments.

Even a cursory glance at the table of contents and the text clearly indicates that this work differs from most elementary level statistical books. The presentation here stresses the planning and interpretation of statistical studies rather than the routine tedious calculations, which turn off a serious nonmathematical professional. Current data analytical tools such as stem-and-leaf plots and box-whisker plots are discussed in this work. The results on regression analysis, analysis of variance, and contingency tables are discussed through Minitab computer outputs. These features, I believe, make this an attractive textbook for a first course in statistics.

The reviews of the first draft of this manuscript helped me tremendously in improving this work, and I am highly obliged to the reviewers for their help. I am greatly indebted to my wife, Damaraju V. Rathnam, and my three children, Damaraju V. Lakshmi, Damaraju Venkatrayudu, and Damaraju Sharada for their constant encouragement and help in completing this project. My special thanks go to Lakshmi for proofreading the manuscript and improving the presentation. I am very thankful to Ms. Janet Evans of Temple University for her patience in skillfully typing the manuscript after deciphering my handwritten notes and to Mr. Richard Lee for his help with some graphics.

I gratefully acknowledge the permission given by the management of Marcel Dekker, Inc., Iowa State University Press, McGraw-Hill, MIT Press, *Time Almanac, 1979,* John Wiley & Sons, *The World Almanac and Book of Facts, 1985;* the editors of *Biometrika and Management Sciences*; and Professors Bechtel, Bishop, Federer, Fienberg, Freund, Holland, Larson, Minton, Steel, and Wiley to reproduce some material from their publications.

Damaraju Raghavarao

Contents

Contents

Exploring
Statistics

1

What Is Meant by Data?

A data set is a statement of factual information that is used to make logical inferences. For all human activities, data are needed. When statistics was in its infancy, statistics was used as a synonym for collection and maintenance of data. With the advancement of statistical knowledge, its current role is to collect, summarize, and analyze the data by mathematical or analytic data models, and to form the necessary inferences from these analyses.

The nature of the data collected, the methods used to collect the data and the type of statistical analysis carried out on the data depend solely on the objectives of the study at hand. The degree of accuracy in interpreting the results also differs from problem to problem. The type of data and the methods that are used by a person in market research to assess the marketability of a proposed new product considerably differ from those of a pharmaceutical company planning to develop a new drug to compete with another available drug.

One should not collect piles and piles of data about every conceivable factor without knowing the type of data needed for meeting the specific objectives of the study. When the collected data are not put to the proper use and interpreted to meet the needs for which they were originally collected, the fundamental purpose for collecting the data is defeated and the situation is comparable to filling a leaky bucket with water.

Many times, the proper type of data may be wrongly interpreted. Federer (1973, p. 35) gives the following satiric example of improper and meaningless statistical use:

1967 Population Balance Sheet or Who's to Do the Work???

Population of United States	198,000,000
People 65 years or over	55,000,000
Balance left to do the work	143,000,000
People 21 years or under	58,000,000
Balance left to do the work	85,000,000
People working for the Government	35,000,000
Balance left to do the work	50,000,000
People on relief and Appalachian Program	24,000,000
Balance left to do the work	26,000,000
People in the Armed Forces	11,000,000
Balance left to do the work	15,000,000
People in City or State Government	12,800,000
Balance left to do the work	2,200,000
Bums and others who never work	2,000,000
Balance left to do the work	200,000
People in hospitals or asylums	126,000
Balance left to do the work	74,000
People in jail	73,998
*Balance left to do the work	2

*Two? Why, that's you, me! Say!! Then you'd better get a wiggle on 'cause I'm getting awfully tired of running this country alone!!   Anonymous

Source: Federer (1973).

Meaningless statistics, shown by the 1967 Population Balance Sheet, arise from overlooking or misinterpreting some or all of the facts. Before proceeding toward an understanding of the foundations of statistics, a clear idea of the types and uses of data is required, and the remainder of this chapter will be devoted to this.

Data is measured in one of the following scales:

1. Nominal Scale
2. Ordinal Scale
3. Interval Scale
4. Ratio Scale

Qualitative data is generally measured on a nominal scale. There is no ordering of the data in this scale; for example, races of people (whites, blacks, Asians, Hispanics); car brands (Cadillacs, Buicks, Chevrolets); types of computers (Apple, Commodore, IBM).

If an order effect can be induced onto a nominal scale, one obtains data on an ordinal or a ranking scale. Movie ratings (excellent, very good, fair, bad, worst); a student's letter grades in a course (A, A−, B+, B, B−, . . .); a person's attitude to a proposition (strongly agree, agree, no opinion, disagree, strongly disagree).

The distance between two values on an ordinal scale is obtained by the interval scale. The ratio between any two intervals is independent of the measurement unit and of the zero point. The unit of measurement and zero point are arbitrary. Quantitative data are usually measured either on an interval or a ratio scale. Some examples of this scale are intelligence quotient (IQ), temperature, and gas mileage.

Finally, an interval scale with a true zero point is called the ratio scale. Some examples of ratio scale data are income, height, weight, and distance.

The quantitative data measured on an interval or ratio scale is either discrete or continuous. The data is discrete if its numerical measurements are countable. If the numerical measurements consist of values over an interval and are uncountable, the data are continuous. If a person gives his height as 72 inches, it is clearly understood that up to the accuracy provided in the measuring device, his height is 72 inches. It could very well be 71.99999 . . . or 72.0010101. . . . Unfortunately, most of the continuous data appear to be discrete due to the limitation of the scale in the measuring device. Observations such as the number of children in a family, the number of home runs made by a baseball player, the number of cars owned by a family, the number of times a housewife visits the grocery store in a month, the number of washing machines sold in a week by a store, the number of daily accidents on a highway, are some examples of discrete data. Such measurements as the heights of people, the blood pressure of people, test scores of students, gas mileage of cars, time to finish a job by workers, percentage of fat in milk, percentage of protein in cereals are some examples of continuous data. Roughly speaking, counting variables provide discrete data and measurement variables give continuous data. Monetary variables are generally considered continuous for the purpose of statistical analysis.

Several examples will now be given to illustrate the uses of data to meet the specific objectives of a study.

Example 1.1.   (Planning). In most countries (including U.S.A.), a census is taken every ten years. The last census year in the US was 1980. During the census, a complete count of the human population in the country will be made. In addition, other useful data will be collected so that the Government can effectively plan the use of its available resources. Examples of useful data are the per capita income over different parts of the country, demand for housing, number of young adults joining the work force, and the distribution of agricultural crops. If data is needed between census years, data will not be collected from all individuals, but will be collected from a smaller portion of the country population using appropriate methods and conclusions will be drawn from that data.

The U.S. Department of Agriculture (USDA) collects information on the acreage and production of crops on a worldwide basis and publishes both domestic and foreign crop reports. Polar orbiting satellites using multispectral scanners make global crop surveys feasible.

Example 1.2.   (Drug Efficiency). When a pharmaceutical company develops a new drug, it must initially conduct a number of experiments on animals to show the drug effect. Such experiments are called preclinical trials. The data collected from these trials will then be submitted to the Food and Drug Administration (FDA) and after getting the necessary approval, clinical trials will be conducted with human beings. Upon satisfactorily establishing the drug efficacy with human data, the company may finally get Government approval to market the new drug.

Example 1.3. (Birthrate and Blackout). To study whether the great New York City blackout on November 9, 1965 created an increase in the birthrate of the city 9 months after the blackout, Izenman and Zabell (1981) used the data of total births in New York City during 1961–66 and the daily birth data from six individual hospitals in the city during August 1966. After appropriate statistical analysis, they concluded that there was no substantial increase in the birthrate resulting from the blackout.

Example 1.4. (Gas and Dust Explosions in Coal Mines). To predict the number of fatalities from gas and dust explosions (Y), as a function of the number of cutting machines in use (X_1), coal output in short tons per labor hour (X_2), the average number of miners working underground (X_3), the number of mobile leading machines in use (X_4), the number of continuous mining machines in use (X_5) and the total annual labor hours of production work (X_6), data for the period 1915 through 1978 was collected. Using that data, Lawrence and Marsh (1984) developed an equation to predict the value of Y given the values of X_1 to X_6 for any year. The equation used to predict one variable given the measurements for one or several other variables is called a regression equation.

Example 1.5. (Elections). In election years it is a common practice to take on-going polls to assess the popularity of candidates. Some of the widely reported, reliable polls are the Gallup, Harris and CBS polls. Given the data obtained from these polls, candidates revise their strategies, change their platforms on issues, or even withdraw from the race. The political future of the candidates mainly depend upon the data from these polls and on the reliability of the data that are collected.

Even on Election Day, the news media take exit polls at key precincts and declare the probable winner before the vote count is completed.

Example 1.6. (Data Affects Data). Unfortunately because the time differences between the East and the West Coasts of the US, while the polls are closed and news media starts predicting the winner on the East Coast the voting process still continues on the west. Human psychology inclines to follow the winner, thus, it is quite likely that the declared results on television could influence the voting pattern on the West Coast. Hence, data obtained and analyzed from the East Coast has the potential for affecting the data yet to come from the West Coast.

This example also underlines a cardinal principle of statistics: one should not plan the statistical analysis or assume the conclusions of a study until all data are collected. An exception to this is the Sequential Analysis procedure where the data are collected in stages. Data should be collected after the objectives are clearly stated and the appropriate method of statistical analysis is decided.

Example 1.7.   (Duodenal Ulcer). Semenya et al. (1983) discussed a clinical study involving patients with duodenal ulcer. A group of 417 patients were included in the study. The procedures involved surgically removing different amounts of the stomach. They are vagotomy and drainage (none), vagotomy and antrectomy (about 25%), vagotomy and hemigastrectomy (about 50%), and gastric resection (about 75%). The number of patients assigned to these procedures were 96, 104, 110, and 107 respectively. The number of patients in each category were then classified according to the dumping syndrome (none, slight, and moderate). The dumping syndrome may be considered as a response variable that has ordinally scaled categories. The surgical procedures have quantitatively scaled categories. Some objectives for such data include comparisons of the procedures for the ordinal distributions of the response variable and the modeling of a relationship for such comparisons with the quantitative structure of the procedures.

A similar type of analysis can be made by taking samples on people in different income brackets and collecting data on their responses toward any proposition measured by an ordinal scale.

Example 1.8.   (Cereals—Their Flavor and Nutritional Value). Bechtel and Wiley (1983) made an interesting study and clustered 13 brands of cereals based on their flavor and nutritional value as given in Figure 1.1. Data were collected on the preferences of 151 male college students for the 13 cereals based on the flavor, which was classified as plain, in-between, and highly flavored, and nutritional value classified as junky, in-between, and nutritious. The data were statistically analyzed to form the indicated cluster pattern.

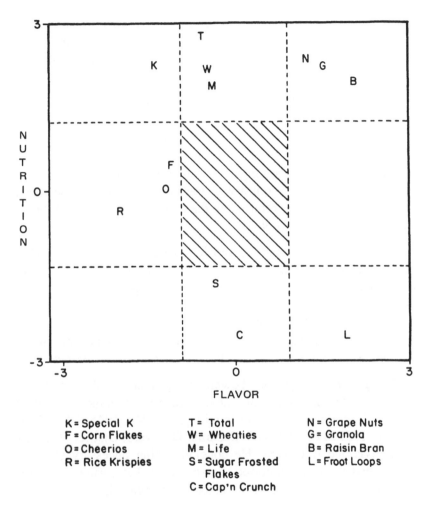

Figure 1.1   Analysis of the flavor and nutritional values of popular cereals. Courtesy of Bechtel and Wiley (1983).

Example 1.9.   (Major League Baseball). The *Baseball Encyclopedia* (1974) provides data on a number of variables: runs scored (*R*), singles (*H1B*), doubles (*H2B*), triples (*H3B*), home runs (*HR*), hits (*H*), total bases (*TB*), walks (*TBB*), hit by pitcher (*HP*), stolen bases (*SB*), caught stealing (*CS*), grounded into double play (*GIDP*), outs (*OUT*), sacrifice hits (*SH*) and sacrifice flies (*SF*). Such data can be used to compare player and team performances. Bennett and Flueck (1983) used such data to develop an expected run production model and recommended that model for an in-depth and precise analysis of teams and players.

Example 1.10.   (Circulation Trial of Bank Notes). Koeze (1979) reported a study in which bank notes printed on different qualities of paper (with or without flax) were examined for the average time until failure. The average time to failure of notes printed on paper without flax was determined to be longer than the notes printed on paper with flax. He observed that, contrary to the prevailing belief, the rate at which the bank notes are returned to the bank is independent of the age and deterioration of the note.

Example 1.11.   (Vitamins and IQ). Lehmann (1975, p. 131) reported a study of Harrell (1943) in which 74 children who were living in an orphanage were divided into 37 matched pairs. One child, randomly selected from each pair, received thiamine while the other, serving as a control, received a placebo. The gains in IQ during 6 weeks of the experiment for 12 of the pairs were analyzed. It was concluded that thiamine had no effect in increasing IQ scores.

Example 1.12.   (Trend Toward Illegitimate Births). Lehmann (1975, p. 295) analyzed vital statistics data concerning the reported number of illegitimate births during the years 1940, 1945, 1947, 1950, 1955 and 1957 in the states of Alabama, Delaware, District of Columbia, Kansas and Louisiana. He concluded after necessary statistical analysis that the data supported the statement that during this period the number of illegitimate births tended to increase in those states.

Example 1.13.   (Forecasting Athletic Records). The problem of forecasting the future records for an athletic event on the basis of the observed past records in that event was considered by Tryfos and Blackmore (1985). By using the data from six actual records of running events from 1968 through 1982, they forecasted the records from 1983 through 1997.

Example 1.14.   (Smoking and Myocardial Infarction). By using the data on myocardial infarction (MI) and cigarette smoking for 393 women in the 35 to 39 years age-group, Denman and Schlesselman

(1983) statistically concluded that 26% of MI are attributable to the smoking of 25 or more cigarettes per day, all other factors being equal.

Example 1.15. (Classification). Rao (1968) considered the data on stature (X_1), sitting height (X_2), nasal depth (X_3), nasal height (X_4), for 3 Indian castes, namely, Brahmins, Artisans and Korwas. He calculated 3 linear functions of X_1 to X_4 known as linear discriminant functions whose numerical values enable to classify an individual into one of these three castes.

The 15 examples just discussed illustrate a microcosm of statistical applications of available or collected data in all walks of life. The statistical analyses used to meet the objectives of these studies differ in complexity. In the following chapters the readers will be introduced to the basic tools of collecting, summarizing and analyzing statistical data.

REFERENCES

Bechtel, G.G. and Wiley, J.B. (1983). Probabilistic Measurement of Attributes: A Logit Analysis by Generalized Least Squares, *Marketing Sci.,* 2, 389–405.

Bennett, J.M. and Flueck, J.A. (1983). Evaluation of Major League Baseball Offensive Performance Models, *The American Statistician,* 37, 76–82.

Denman, D.W. III, and Schlesselman, J.J. (1983). Interval Estimation of the Attributable Risk for Multiple Exposure Levels in Case-Control Studies, *Biometrics,* 39, 185–192.

Federer, W.T. (1973). *Statistics and Society,* Marcel Dekker, New York.

Harrell, R.F. (1943). Effect of Added Thiamine on Learning, Contrib. Educ. No. 877, Teacher's College, Columbia University.

Izenman, A.J. and Zabell, S.L. (1981). Babies and the Blackout: The Genesis of a Misconception, *Soc. Sci. Res.,* 10, 282–299.

Koeze, P. (1979). An Accurate Statistical Estimation of the Life-Length of f-100 Banknotes, *Int. Statist. Rev.,* 47, 283–297.

Lawrence, K.D. and Marsh, L.C. (1984). Robust Ridge Estimation Methods for Predicting U.S. Coal Mining Fatalities, *Comm. Statistic,* 13, 139–149.

Lehmann, E.L. (1975). *Nonparametrics: Statistical Methods Based on Ranks,* Holden-Day, Oakland, Calif.

Rao, C.R. (1948). The Utilization of Multiple Measurements in Problems of Biological Classification, *J. Roy. Statist. Soc.,* 10B, 159–193.

Semenya, K.A., Koch, G.G., Stokes, M.E., and Forthofer, R.N. (1983). Linear Model Methods for Some Rank Function Analyses of Ordinal Categorical Data, *Comm. Statist.,* 12, 1277–1298.

The Baseball Encyclopedia (1974). Macmillan, New York.

Tryfos, P. and Blackmore, R. (1985). Forecasting Records, *J. Am. Statist. Assoc.,* 80, 46–50.

2

How to Collect Data

Chapter 1 indicated that the objectives of a study are met by analyzing appropriate data. The required data for a study can be obtained by surveys, experiments, or published records. The survey method of collecting data differs considerably from the experimental method. In the experimental method, the necessary data will be generated by a planned experiment; while in the survey method, the pertinent data are available at the source and the investigator collects suitable data by appropriate methods.

The fundamental concepts behind any statistical investigation are the notions of population and sample. The purpose of this chapter is to explain these ideas and introduce the reader to the art of collecting data.

POPULATION AND SAMPLES

A population consists of all the observations that one is concerned with in a study. The interest is not that of a complete knowledge of each population observation, but in knowing some of the values that characterize the entire population, such as its average. A statistical population is different from human or animal populations. A *sample* is a subset of the known population and is used to draw inferences about that population, from which the sample is taken. The following examples illustrate the population and sample concepts.

Example 2.1. (Length of Life of Washing Machines). Suppose one is interested in knowing the average length of life of brand A washing machines. Here the population is the lengths of life of each of the brand A washing machines. Since it is impractical to collect data on the whole population, one obtains data for a predetermined number of randomly selected machines. Conclusions concerning the length of life of all machines of brand A will then be made using the sample.

Example 2.2. (Artificial Heart). Suppose one needs to know the life extension provided by using an artificial heart in patients over 50 years of age who have severe heart problems. The population here is the patients whose life extension will be provided by this operation. Clearly, it is not possible to obtain this data. One may use the data on the patients who received artificial hearts to draw inferences about the performance of the artificial heart. The data collected on the operated patients are the sample. When data are collected at any point of time, these data pose an interesting and peculiar problem. People who received artificial hearts 10 years ago and are living at the time of the study have extended their lives by 10^+ years with this device. The data may thus have numerical values with a $+$ sign. This type of data is used in survival analysis problems.

Example 2.3. (Smokers). In a study to determine the percentage of smokers in a city, one faces the difficulty of determining who these people are. One needs an operational definition of a smoker. An operational definition of such a person may be: "a person who smoked at least once before," "a person who smokes at least one cigarette every day," "a person who smokes at least one pack of cigarettes a day." The definition should meet the technical objectives of the study. Subject to the definition given for smokers, each person in the city can be classified as a smoker or a nonsmoker. The population here is the *yes* or *no* categorical response of each individual in that city. The number of people who smoke divided by the number of people in the city is the proportion of smokers in that city and the objective of a study is to determine that value. For this purpose, the investigator interviews a predetermined number of randomly selected individuals and determines whether each is a smoker or a nonsmoker. The *yes* or *no* categorical response given by each interviewed person is the sample data. The inferences about the proportion of smokers in the city will then be made using the sample data collected from the interviews.

Example 2.4. (Grade-Point Average). Suppose you are interested in predicting the Grade-Point Average (GPA) of students in a semester. Clearly the admission test score (X_1), the number of credits registered in that semester (X_2), and the number of hours spent on the job (X_3) are some of the variables which affect the GPA of a student. Thus, one expects to develop a functional form to predict the GPA (Y) of a student given the student's profile values of X_1, X_2, and X_3. Here the population data is Y, X_1, X_2 and X_3 for each of the past, present, and future students whose GPA is being predicted. The data for Y, X_1, X_2 and X_3 on a subset of students will be considered for statistical analysis, and this makes up the sample data. By using this sample data, the prediction equation will be developed and used to formulate the GPA.

Example 2.5. (Speed-Reading Course). Let the objective of a study be the examination of the effectiveness of a speed-reading course. The effectiveness can be measured in two ways:

1. The percentage of individuals who increase their reading speed after taking the course,
2. The average increased number of words/minute reading after the program

To evaluate the effectiveness in either case, one needs data indicating the increased reading speed after taking the course. The population consists of the increased speeds in reading for all people who take the course. It is not feasible to obtain such data. Instead, randomly selected volunteers will be used as a sample. The reading speed before the course and at the completion of the course will be taken for each volunteer. A negative value for the increased reading speed indicates a decreased reading speed. The sample data is the increased reading

speed of the volunteers used in the study. Given this sample, conclusions about the effectiveness of the course can be drawn.

The important features of collecting data by survey method will be
considered in the next section.

SURVEY METHOD

When the necessary population is already available and when there is
no need to generate the data, survey methods are appropriate to collect
the necessary data. The data can be collected by personal interviews or
mailed questionnaires. The ultimate unit for which the data is collected
by a survey method is called a *sampling unit* (SU). An SU in a survey
may be a person, a family, a country, a company, an animal, or an
object. Let the population consist of N SUs giving N observations for
the population data. N is called the population size. A census is the
survey of N population SUs and gathering all the data. As noted
before, most of the countries conduct a census of their citizens every
ten years. From the examples discussed in the last section, it is clear
that one may or may not be able to conduct a census for any specific
problem. Even when taking a census is possible, it is prohibitive in
some cases because of the expenditure and time needed in doing it.
When the census is done and data analyzed, the problem itself might
have been solved or lost its importance and the results become ob-
solete. A quicker understanding of the population data can be reached
at a lower cost by surveying a portion of the population SUs. The SUs
used to collect the data are called sample SUs and the data collected
thereby is called a sample. The number of SUs in the sample is usually
designated by n, the *sample size*. The validity and accuracy of the
inferences drawn from the sample depend on the value of n and the
method of drawing the sample. In every problem there is a minimum n
that provides the estimates to the desired level of accuracy and the
mathematical aspects of this will be discussed in the Appendix of
Chapter 4. By obtaining data on more than the minimum n of sample
SUs, one may achieve more accuracy at a higher cost. The minimum
value of n is used in practice.

The method of organizing and collecting data from sample SUs is
called a sample survey. In planning a sample survey, the following
aspects (cf. Ford and Tortora, 1978) must be considered by the inves-

tigator so that meaningful interpretations can be made from the collected data:

Objectives of the Survey

The objectives of the survey must be clearly decided before planning the survey. One should not go on a fishing expedition to collect all types of data from SUs with the hope of devising objectives after examining the body of the data. A method of choosing sample SUs for one objective may be useless and may give meaningless conclusions for another objective of the study.

Example 2.6. (First Statistics Course). Consider a study involving college students with the following two objectives:

1. to determine the average percentage points obtained in the first statistics course
2. to determine the percentage of students who enjoy the first statistics course.

Objective (1) can be met by collecting the data of the percentage of points for a sample of students providing a cross section of the students of that college. Such a sample is called a simple random sample. Those sample SUs forming a simple random sample may not provide appropriate data to answer objective (2). It is reasonable to expect that students getting a better grade in a course usually like it. As such, it is desirable to group the students based on their scholastic performance and then take a simple random sample of SUs from each group. The sample so obtained is called a stratified random sample.

This illustration clearly focuses the importance of specifying the objectives of a study to decide the appropriate method of choosing sample SUs. In complex surveys with multiple objectives, some compromises have to be made.

Population to Be Covered

That population for which the investigator aims to draw inferences is known as the *target population*. If because of limitations on the study and the nature of the data, this aim can not be fulfilled the investigator must be content in drawing his conclusions from a subset of the target population. The population from which the sample is actually taken and is representative of it, is called the *sampled population*. All inferences based on a sample survey are valid for the sampled population, but not necessarily the target population. A lack of understanding of the difference between target and sampled populations may lead to misinterpretation of the results. The following examples indicate the differences between these two populations.

Example 2.7. (1936 Presidential Election). The *Literary Digest* predicted, based on sample data in which the sample SUs were selected from people listed in telephone directories and magazine subscription rolls, that Republican candidate Alfred M. Landon would defeat Democratic President Franklin D. Roosevelt. The results were just the opposite and President Roosevelt had a landslide victory. The *Literary Digest* committed the mistake of drawing an inference about the target population of all voters from the sampled population who were drawn from the society of more affluent voters who had telephones or subscribed to several magazines.

Example 2.8. (Sales Promotion). One may sometimes get a promotional advertisement from a car dealer stating that a gift will be given if the customer will test-drive their new model car. Usually such offers will be sent through letters to their past customers. The intention of the dealership is to promote the sales of their new model cars to all prospective buyers. However, they are addressing themselves only to past customers. If 10% of the people to whom the promotion letters are sent avail themselves of the offer, it is wrong to conclude that 10% of the prospective buyers test-drove the car. It is only true that 10% of the previous customers of that dealer test drove the new car. In this example the target population SUs are all prospective buyers and the sampled population SUs are the previous customers of that dealership.

Frame

As a guide to the investigator for locating the population and sample SUs, one may need an appropriate listing of all population SUs; such a listing is called a *frame*. A frame may be a list of voters, road maps, telephone directories, customer invoices, list of taxpayers, and so on. If a frame is not available to meet the objectives of a particular study, one must develop an appropriate one to suit specific requirements. The frame holds the key to bridging the gap between target and sampled populations. A poorly constructed frame can easily make an expensive study worthless.

Sample Size and Sample Selection

The main ingredients for a well-finished sample survey are the appropriate choice of sample size and the method of drawing the sample SUs. The importance of the sample size was indicated at the beginning of this section and can be easily seen in the following example:

Example 2.9. (Fuel Consumption in Cars). Suppose a large-fleet car buyer wants to determine the average gas mileage for a given brand of car with specific equipment options. If all cars have the same fuel consumption, by testing even one car the buyer will be able to determine exactly the average gas mileage of all of the cars. It is known that no two identically equipped same brand cars give the same fuel consumption. One will be able to determine the average fuel consumption by test driving several cars. The average fuel consumption based on 100 cars is closer to the actual average of all cars than that based on 10 cars. Similarly, the average fuel consumption based on 1000 cars is much closer to the actual average of all cars than that based on 100 cars. However, it is more expensive to test drive 1000 cars than 100 cars. This example indicates that small samples increase the error in estimating the values of interest while large samples increase the cost of the survey. In determining the sample size a balance had to be struck between the accuracy in estimation and the cost.

The problems of determining the appropriate sample size and the method of drawing sample SUs are interrelated. One can determine the appropriate sample size to give the desired accuracy, subject to some random chance element if the method of drawing sample SUs is decided. On the contrary, if there is a limitation on the sample size due to some restraints in the study, one can decide accordingly the method of selecting sample SUs to meet the goals.

There are several methods of drawing the sample SUs in a survey. The simplest of them are (1) Simple Random Sampling, (2) Stratified Sampling, (3) Systematic Sampling, and (4) Cluster Sampling. A simple discussion of these methods and the scenarios in which they are valuable follows.

(1) Simple Random Sampling

In simple random sampling each population SU has an equal chance of being included in the sample. No SU is either favored or purposefully eliminated from inclusion in the sample. This method is useful when the data on population SUs are homogeneous and there is reason to believe that the data are not affected by some extraneous classification of population SUs.

Given the frame, the population SUs will be numbered from 1 to N. Random numbers will be generated on a computer where each number is between 1 and N, and those units corresponding to the numbers selected by the random numbers are surveyed, to collect the data. In the process the same SU may be picked more than once. The method that allows the use of repeated SUs in the survey is called simple random sample with replacement, while the method not allowing the use of repeated SUs in the survey is called a simple random sampling without replacement. While doing the sampling without replacement, the repetitions of the already drawn random numbers will be ignored. Random numbers are drawn until the required sample size is reached. Some inferences based on data collected from a simple random sample with replacement will be considered in Chapter 4.

(2) Stratified Sampling

If the data are heterogeneous and if the investigator is able to divide the population SUs into groups (called strata) such that the data in each strata are reasonably homogeneous, this method is applied. Simple random samples of appropriate sizes are selected from each strata. By using stratified sampling, one gets data representing each of the groups in the population.

(3) Systematic Sampling

Let N and n be the population and sample sizes and let N/n be rounded to the nearest integer k. A random number between 1 and k will be selected and from the given frame, the SU corresponding to the randomly selected number and every kth SU from there on will be included in the sample SUs. In certain settings this method provides a convenient method of choosing the sample SUs to collect the data.

(4) Cluster Sampling

Another useful technique to collect the data is to sample in clusters or groups rather than the individual SUs. When the units are spread out over a large geographical area, selecting a sample of SUs becomes very expensive. Here, one may group the SUs to form clusters of SUs, select a sample of clusters and collect the data on each unit in each of the selected clusters. It is also possible to subsample the clusters.

The difference between stratified and cluster sampling is worth noting. In both schemes the population is divided into groups. While the data in each group are nearly homogeneous in stratified sampling, it is not necessarily so, in cluster sampling. In stratified sampling, data are collected from each strata, whereas in cluster sampling a sample of clusters are selected for collecting data.

Examples will now be given to demonstrate the use of these four methods of choosing sample SUs.

Example 2.10. (Traffic Volume). Suppose one is interested in determining the average traffic volume at a bus stop during the peak morning hour. The volume on all the days will be more or less the same for all weekdays and hence there is no need for a stratified sample. There are also no natural clusters of weekdays and hence cluster sampling has no advantage. One may pick a day of the week randomly and collect the data on that day of every week using systematic sampling. The use of this procedure risks acquiring a systematic error if the randomly selected day has uniformly lower or higher traffic volumes. The ideal method of collecting the data in this problem is simple random sampling without replacement. A sample of days can be randomly selected and the traffic volume on those days can be determined by the investigator.

Example 2.11. (Iranscam). Suppose a political scientist wants to estimate the popularity of President Reagan among Philadelphia voters after his televison address on March 4, 1987 responding to the Tower Commission Report. Since the voter's party affiliation is likely to bias the popularity, it is appropriate to take a stratified sample of voters using two strata, Democrats and Republicans. By taking a simple random sample, the investigator runs into the risk of selecting all Democrats or Republicans, biasing the study.

Example 2.12. (Market Survey in a Mall). When one visits suburban shopping malls, it is common to see interviewers soliciting data on market surveys. The appropriate method of collecting data in such a setting is systematic sampling. The only frame available to the investigator is the order in which people enter the mall. Because this frame cannot be listed, it is not possible to use simple random, stratified, and cluster sampling procedures to collect the data. One can randomly choose a person for data collection and then follow up by collecting data on every kth person entering the mall for a suitable k.

Example 2.13. (Job Satisfaction). Suppose a national chain store is interested in a personal interview survey of its employees concerning their job satisfaction. It is very expensive to collect data by simple random sampling because interviews have to be conducted all over the country. There is no need to stratify the population because the job satisfaction of all such employees is expected to be more or less uniform. Systematic sampling is also uneconomical. The useful data collection method here is cluster sampling using cities as clusters. One can randomly select a limited number of cities from all cities in which the store has branches and collect data on all or a sample of employees in each of the selected cities.

Example 2.14. (Acceptance Sampling). When a large consignment of items is ordered by a company, and if the items are packed in cartons, a commonly adopted procedure is to quality check all items in a sample of randomly selected cartons. If the number of defectives in the examined sample is less than some predetermined number, the consignment is accepted. If the number of defectives exceed another predetermined number, the whole consignment is rejected. If the number of defectives is between the rejection and acceptance limits, the whole consignment is examined for accepting or rejecting the consignment. This method of data collection is easily verifiable as cluster sampling.

The Information to Be Collected

One should properly plan and collect exactly enough information to meet the goals of the study. Inadequate information will not enable the investigator to address the issues of the study. Conversely, if data are collected on unnecessary items, the collection and analysis of data will not be completed on time and the objectives of the investigation may be outdated by the time the answers are available.

Because of the sensitivity of the questions and the legal implications, respondents may sometimes not cooperate to give data on sensitive questions. The nature of sensitive questions differ from community to community. Questions which are considered sensitive in one community are freely and openly answered in other communities. In the last 20 years, substantial research was done in eliciting answers to sensitive questions following the lead of Warner (1965). All these techniques are based on the premise that the respondent will give a correct answer if the identity of the person is protected; these are called randomized response procedures. Suppose, in a survey, the investigator wants to ask a woman whether she has had an abortion. Because of the delicateness of the question and because of people's reluctance to divulge secrets to strangers, the respondent may, to say the least, hesitate to answer the question. A simple randomized response procedure is to form two questions:

1. Have you ever had an abortion?
2. Are you a Capricorn?

The respondent is given a die telling her that she should throw the die without the investigator's knowledge and answer (1) if the die face shows 1 or 2 and answer (2) otherwise. She should be clearly told that it is not necessary for her to inform whether she is answering (1) or (2) and she should only answer yes or no for the question she has selected based on the throw of the die. By mathematical formulas the proportion of women, who have had an abortion, can be estimated using the proportion of *yes* answers collected by the investigator.

A method provided by Smith, Federer, and Raghavarao (1974) will enable one to use several sensitive questions in a single survey. Their illustration of the technique is discussed in the following example:

Example 2.15.   (Cheating, Stealing and Smoking). Smith, Federer and Raghavarao (1974) wanted to estimate the proportion of students in a classroom at Cornell University who cheated on the test, who stole at least 5.00 dollars, and who smoked marijuana. In addition to these three sensitive questions, four more nonsensitive (at least partly) questions are included in the study. Because the response to each of the seven questions is a categorical *yes* or *no*, the answers are scaled. The seven questions used by them and the scales for the responses are

1.   Are you under 21 years of age?
     Yes (0)   No (1)
2.   Did you cheat in any way on the Stat 200 prelim that you took last
     week?
     Yes (2)   No (3)
3.   In general, are you happy with your decision to come to Cornell?
     Yes (1)   No (2)
4.   While at Cornell, have you ever stolen money or any other article
     worth over 5.00 dollars from a roommate, friend, employer, or
     any one else?
     Yes (3)   No (2)
5.   Does your parent earn more than 25,000 dollars a year?
     Yes (0)   No (1)
6.   Have you smoked any marijuana during the past 2 weeks?
     Yes (2)   No (3)
7.   Are you enrolled in the College of Agriculture and Life Sciences?
     Yes (1)   No (0).

The scores for *yes, no* answers for all questions can be given 0 or 1
also, but it is felt that more confidence could be created among the
respondents by arbitrary scoring. The difference in yes – no scores
must, however, be 1. The seven questions were then grouped into
seven sets as follows:

$$S_1 = \{Q_1, Q_5, Q_6\},\ S_2 = \{Q_4, Q_5, Q_7\},$$
$$S_3 = \{Q_3, Q_4, Q_6\},\ S_4 = \{Q_1, Q_3, Q_7\},$$
$$S_5 = \{Q_2, Q_6, Q_7\},\ S_6 = \{Q_1, Q_2, Q_4\},$$
$$S_7 = \{Q_2, Q_3, Q_5\}.$$

It is preferable that each set contains both sensitive and nonsensitive
questions. The above formed sets are known in the literature as a
balanced incomplete block design. Eighty-four respondents were used
in the survey and they were randomly divided into seven groups of 12
students each. Each of the 12 students in the ith group were asked to
give the value to the sum of their answers to the three questions of the
ith set. Hence no student answers all questions and every student
answers only three questions, not individually, but providing a total
value to his/her answers of the 3 questions. For example, a student

answering set 3 has responses, no, yes and, no to questions Q_3, Q_4, and Q_6 respectively, simply reports 6 to the interviewer without divulging the true answers to each question. An answer of 6 in that set can also be seen to arise if the respondent's actual status to questions 3, 4, and 6 are respectively yes, no, and no, thereby protecting the anonymity of the respondent's answers. The data was then analyzed to find the estimates of the percentage of students who cheated on the exam, who stole money and who smoked marijuana, in addition to the percentage of students who are under 21, who are happy to join Cornell, whose parents earn more than 25,000 dollars a year and who were enrolled in the College of Agriculture and Life Sciences. One of the drawbacks of this procedure is that the choices of the respondent is revealed when the reported total value is the minimum or maximum value for that set total.

The previous discussion indicates that there are special tools available when the interviewer wants to acquire information on sensitive questions, by protecting the respondent's identity relative to the true status of answers to the questions.

Method of Collecting Information

Information can be collected either by mailing questionnaires or by personal interviews using schedules. Both of these methods have advantages as well as disadvantages. It is less expensive to do a mail survey, but there is an associated risk of a higher nonresponse rate. Furthermore, the questionnaires should be very carefully and clearly worded such that no confusion on the part of the respondent will be created while answering it. The personal touch included in the interview method provides less nonresponse although it is expensive to train the interviewers and pay their salaries.

Preparation of Questionnaires and Schedules

The instrument used in a mail survey is called a *questionnaire* and the one used in a personal interview is called a *schedule*. The investigators should use their maximum skills to draw up a questionnaire or schedule that includes the right type of questions required to meet the specific objectives of the study and the correct order to ask them. The questions should be simple and unambiguous leaving no room for the respondent to interpret them to their liking. Filtered questions can be used where needed. Suppose one wants to know the demand for different brands of cigarettes. One should not directly ask "What brand of cigarettes do you smoke?" A filtered question is first used by asking "Do you smoke?" and is continued to the next part of the question, "If your answer is *yes,* what brand of cigarettes do you smoke?" It is a good practice for the investigator to pretest the questionnaire or schedule on a group of people similar to that for whom the survey is directed. Given the pretest experience, one can decide to use the same instrument for the main survey, or modify it appropriately.

Auxiliary Information

In some survey situations, the available or collected information on an extraneous (or auxiliary) variable will enable the investigator to estimate the values of interest more accurately. Such information will also be useful to stratify the population for a survey. Use of auxiliary information is discussed in the following examples:

Example 2.16. (Recall Petition). Suppose signatures are collected at several places in a city on a recall petition of a public official. The total number of signatures collected can be easily determined. However, all those signatures may not be authentic signatures of the voters in that city, and some people may have even signed the petition more than once. It is a difficult task to check the truckloads of signature sheets and to determine the exact count of authentic signatures. An alternative procedure using the idea of auxiliary information is to select a simple random sample of sheets, check the signatures for their accuracy and determine the percentage of useful signatures. Because the total number of signatures on all of the sheets can be counted, one can then easily estimate the number of valid signatures on the petition. It should be noted that the duplicate signatures on the petition may create some problem and this can be avoided if the duplicate signatures are considered invalid. Otherwise, more specialized techniques may have to be developed to answer the issue.

Example 2.17. (Population Estimation During Midcensus Periods). Suppose one wants to estimate the population in a certain geographic region in 1987. The region consists of several counties and the population in those counties was known from 1980 census data. A simple random sample of counties is selected and the population in those counties can be determined by a sample survey. Given the data of the selected counties one can determine the proportion of 1987 population using the 1980 population as the base. By using this proportion and the census data of 1980, one can easily estimate the 1987 population for the whole region.

Training of Interviewers and Interviewer Bias

It is highly essential that all interviewers collecting data should be highly and identically trained so that they collect similar data without using their personal judgment. Providing good training program for the interviewers is essential for conducting a survey. Whenever possible, it is desirable to assign overlapping SUs to interviewers for collecting data so that the interviewer's bias can be examined.

Inspection of Returns

Completed returns of questionnaires or schedules should be scrutinized for possible omissions, errors, and impossible observations. A response of 1500 pounds as the weight of an individual shown on a return is obviously an error possibly resulting from the inadvertent addition of an extra zero by the interviewer or respondent. When such discrepancies are observed, it is better to have the data reexamined rather than make corrections and hoping for the best. Whenever possible, the data on the returns should be compared with other available data.

Errors

Two types of errors occur in almost all sample surveys. Even the best efforts on the part of the investigator cannot completely eliminate these errors. At best, they can only be minimized. The first type of error is called the sampling error and is uncontrollable. When one is planning to draw valid conclusions from population data by using only a small portion of it, it is conceivable that the entire population does not possess the same response and we might have picked up all high observations or all low observations unintentionally. In a survey to determine the average family income, one may, purely by chance select all wealthy families or all poor families for the survey and create a large error in estimation. This error is unintentional and creeps in because of the way the sample is selected. This error can be reasonably estimated and one can estimate the population values with a margin of error. During an election year, it can be noticed that television newscasters summarizing election polls remark that candidate A has 48% of the votes with a margin of error of 3%. The statement implies that candidate A is expected to get 48% of the votes from the electorate. However, one should not be surprised if the actual percentage is between $48 - 3 = 45\%$ to $48 + 3 = 51\%$. The 3% error is a measure of the sampling error.

The second type of error is the nonsampling error and the investigator can take at least some steps to minimize this error. Nonsampling errors occur because the respondents may wilfully give wrong answers or the investigator may not be able to collect the data from some sample SUs. If the response errors are random, the total response bias is negligible and the survey results are acceptable. Intentional response errors can be minimized by adequate scrutiny of the returns and using randomized response procedures to provide confidentiality of the responses. When the interviewer is unable to collect data from a SU, it might usually be due to one of the following reasons:

Noncoverage
The interviewer may not be able to locate the SU, or there is no proper transportation to visit the selected SU.

Not-at-homes
When both men and women work, it is possible that the interviewer may not find the respondents at home when he or she visits them. Standard sampling theory textbooks provide methods to cope with this situation and the interviewer may repeatedly try to approach the SU and collect the data. However, great caution must be exercised when using data from repeated calls as is evident from the following example.

Example 2.18. (Missing children). A survey was conducted to estimate the number of children per family. Based on the data collected from the first visits, the average number of children was estimated as 3.2. Because there were many not-at-homes in the first try, the investigator made a second try to catch some of the not-at-homes and estimated 2.5 children per family. A third try estimated this number at 2.4. The decrease in the estimated number of children per family should not cause us any concern because the no-children-families became non-responding units in the earlier part of the survey and due to intensive methods of locating the SUs, they entered the survey in the later months, thereby decreasing the average number of children per family.

Unable to Answer
The respondent may not have the answer with which to provide the interviewer.

Hard Core
The hard core are the people who refuse to be interviewed in spite of any kind of persuasion from the interviewers and they should be ignored.

When considering nonresponse errors, the investigator must entertain the possibility that there may be a significant difference in the response between responding SUs and nonresponding SUs.

From this discussion it is clear that a sample survey is done when the population is available. The necessary data is collected by appropriate methods from the population. In situations where the population is unknown, data has to be created by conducting suitable experiments. To generate the right kind of data without biasing the inferences, the experimenter has to follow some guiding principles in planning and conducting experiments. They will be discussed in the next section.

EXPERIMENTAL METHOD

The branch of statistics devoted to the planning and analysis of experiments is known in the literature as *statistical design and analysis of experiments*. In social sciences this is referred to as *research design*. Experimental designs, which originated in agricultural experiments, slowly found their way into experiments in different disciplines. It will be no surprise to find the use of a statistical design in a market research problem to study the effects of different marketing strategies. However, due to the initiation of this subject in agricultural experiments, the terminology of plots, blocks, and treatments are generally used in describing these designs. An ultimate unit used in connection with an experiment is called an *experimental unit* (EU). In clinical experiments dealing with humans, the EU is an individual person. In preclinical trials dealing with animals, the EU is an animal. In agricultural experiments the EU is a piece of land of given dimensions, the dimensions being determined by a method called uniformity trials. In industrial experiments the EU is a piece of machinery. In a marketing experiment the EU is a market. The EU for any experiment can be determined based on its nature.

The treatments are the procedures, practices, medicines, or fertilizers introduced into the experiment by the investigator. While the experimenter is interested in studying the effect of treatments, usually inferences of individual treatment effects cannot be made. Conclusions can be drawn on comparison of effects involving two or more treatments. The treatments in an experiment may be unrelated ones, may be different levels of a factor, may be all combinations of different levels of several factors, or may even be a small fraction of all combinations of the levels of several factors. The choice of appropriate treatments for an experiment is called a *treatment design*. In an agricultural experiment, one may be interested in determining the yields of five different varieties of wheat crop and in that experiment there are five different treatments. Suppose a doctor is interested in testing the effect of Diabenese for the control of diabetes. He may use different doses of Diabenese such as 0 mg., 250 mg., 500 mg., and 750 mg. in an experiment, and study the effect of these doses in the control of the diabetic condition. In a study to determine the effect of humid conditions on the force required to pull apart pieces of glued plastic, there are two factors in the experiment: type of plastic and humidity level. One uses different humidity levels such as 40% and 80% and uses different plastic types such as Plexiglass and Vinylite. The treatment design then consists of 4 treatments as follows:

Plexiglass at 40% humidity
Plexiglass at 80% humidity
Vinylite at 40% humidity
Vinylite at 80% humidity

A treatment design in which the treatments are combinations of the levels of several factors are called factorial experiments. In industrial experiments a large number of variables (or factors) affect the response of the study variable and it is practically impossible to run the experiment using combinations of different settings of all the variables under study. In such experiments a suitable fraction of all treatment combinations will be used as treatments in the experiment; they are called fractional factorial experiments.

It is often difficult to obtain identical or homogeneous EUs for conducting an experiment. One then groups the EUs such that EUs within each group are nearly homogeneous, whereas EUs in different groups may or may not be homogeneous. The groups so formed are called blocks. It is recommended that in an experiment no block should consist of more than 10 EUs as the experimental error (to be discussed in Chapter 6) is likely to increase with larger blocks. In agricultural experiments contiguous plots are combined to form the blocks.

Depending on the nature of the experiment, one may give only one treatment to each EU, or may give treatments in a sequence over different periods of time. It is also possible to treat an EU at the beginning of the experiment and take observations over different periods. Designs in which observations are taken over several periods are known in the literature as repeated measurement designs and EUs in such experiments are usually called subjects. If each subject receives a sequence of treatments, the associated designs are called crossover designs. Crossover designs, whenever feasible, are economical and provide more accurate comparisons of treatment effects.

Three basic principles for the experimental method of collecting data are: (1) *replication,* (2) *randomization,* and (3) *local control.* To obtain good results from the conducted experiments, the investigator must carefully consider these principles in planning the study.

Replication

Replication is the repeated use of a treatment during a study. It is essential to make the study feasible. Though the experimenters think that their experiments are well controlled, the variability present in such trials can be noted by replicating even a single treatment. The observations arising from EUs receiving the same treatment will be considerably different! By increasing replications of treatments an experimenter will be able to detect the differences between the treatment effects, whenever such differences exist. With fewer replications, even a very well-organized experiment will become useless. Attention should be paid to use an adequate number of replications to make the study meaningful (see Appendix). The next example shows the need for replications in an experiment.

Example 2.19. (Why Replicate?). Consider an experiment in which two weight reducing diets A and B are compared. Suppose two people of the same sex, age, weight, and health condition are used as EUs one receiving diet A and another diet B. Suppose that at the end of a stipulated period, say 3 weeks, the weight losses from diets A and B are observed as 5 and 7 pounds. Noting that all identical types of people receiving the same diet do not show the same weight loss, it is difficult to determine whether the loss of 5 and 7 lb could have arisen purely by chance or whether the programs are really different. Unless the experimenter has knowledge of the extent of variability that can arise within each of the diets, it is difficult to discard the random chance theory and to conclude that the two diets produce different weight losses. To study the extent of variability that can arise from each of the diets, it is imperative to use each of the diets with several persons and then, given the data one can measure the variability between the groups and compare it with the variability within the groups. If it becomes evident through statistical theory that the observed variability between groups cannot arise by random chance, then the two diets will be assumed to produce different weight losses. A single replication for each diet will in no way help the experimenter to conclude one way or other, the equality of the diets' effects in weight reduction.

Randomization

The physical process of applying treatments to EU is accomplished through the mechanism of randomization. It makes the statistical analysis valid by making it appropriate to analyze the data as though the assumptions made for the analysis are true. Usually the experiments are expected to be under complete statistical control. Occasionally, minor departures may occur and the randomization process reasonably balances out such minor errors and provides valid analysis. It is akin to insurance to protect the experiments, when minor disturbances occur in conducting them. It is not practicable nor convenient to use the device of randomization in all experiments. However, it is a wise policy to use it whenever feasible. Some experimenters and statisticians believe that without randomization the inferences drawn from an experiment are not valid. The following example will convince the reader of the utility of randomization.

Example 2.20. (Why Randomize?). The Police Department is interested in determining which of the two known crime prevention methods is more effective in a city. If the city is divided into small segments and if method A is used first followed by method B in each area, it is likely that more criminals will be caught by method A, because all of the criminals in the area might have been caught in the initial stage of the experiment and no criminal is left to become a statistic for the second method. Furthermore, it is also likely that method A might have scared the criminals and it takes time for the criminal mind to return to normalcy. However, if in each of the city segments method A or B is randomly selected for the first application followed by the alternative method, the real difference between methods A and B can be determined without favoring method A and prejudicing method B or vice versa. There is still a slight possibility that the random assignment of initial and final methods to each area may still yield method A as the first choice in each part of the city and such a possibility is beyond the experimenter's control. Another useful method is to choose some districts randomly and assign method A first followed by method B, while the rest of the districts are given method B first and method A later. This discussion opens up the possibility that the experimenter has several possible choices for planning the experiment and a careful choice of the appropriate method of conducting it, is in order to get valid conclusions. This aspect of the planning of the statistical designs will be discussed later.

Local Control

Any method used in the experiment to increase the chances of detecting even small differences in the treatment effects can be called a local control. This includes the blocking of EUs, creating homogeneous blocks, use of auxiliary information in covariance analysis, and use of control treatments with no effects. The control treatment used in pharmaceutical experiments is a placebo. A placebo looks like the treatments used in the experiment, but has no therapeutic value. In clinical trials a placebo is used to offset the psychological belief of some patients who think that any drug alleviates the symptoms of the disease and cures them. Such a feeling is understandable if one observes small children, who when hurt, run to their elders for a Band-Aid even when it is not needed, with the hope that it is the only solution. The use of a control treatment in an experiment serves two purposes. First, it provides a natural measure of the extent of variation present in EUs under an ideal no-treatment condition. Second, the experimenter cannot measure the effect of a treatment unless a standard is available and the control treatment serves this purpose.

The plans indicating the assignment of treatments to EUs are called *experimental designs*. There are several experimental designs available in the literature. Most of the commonly used, as well as other complicated designs, were discussed by this author (1971) and their statistical analysis was given by Cochran and Cox (1957). Some simple and widely used designs and their applications are now discussed.

Completely Randomized Design (CRD)

In this setting all EUs are nearly homogeneous with respect to all sources of extraneous variations. The EUs will be randomly divided into as many groups as the number of treatments used in the experiment. Each EU of the ith group will be treated with the ith treatment and data will be collected at the end of the experiment. This design has wide flexibility. It can be used with any number of treatments and any number of replications per treatment. Thus it is not necessary to divide the EUs into groups of equal size. Even if the data is lost for some EUs, which does occur sometimes during the course of the experiment for reasons beyond the control of the investigator, the statistical analysis remains quite simple. The main difficulty in the implementation of this design is the selection of the homogeneous EUs needed for the experiment. This design is generally used in laboratory experiments and agricultural experiments conducted in pots by using homogeneous soil for growing the crop, where a more controlled environment prevails.

Randomized Block Design (RBD)

When homogeneous EUs are not available, or when the experimenter wants to study the treatment effects over different levels of a second factor, this design is used. As noted before, the EUs will be grouped into blocks, where each block consists of homogeneous EUs. This design is applied when the number of treatments, preferably, is fewer than 10 and usually not more than 16. The treatments will be randomly assigned to the EUs in each block using different randomizations from block to block. The EUs of the same block will be similarly exposed to other environmental variables excepting for the treatments used. The blocking used in this design will improve the efficiency of the experiment because the experimenter can compare the treatment effects on similar EUs and the conclusions have a wider inductive basis because they are obtained in blocks under different conditions. It is also to be noted that the blocks need not be at the same location. Missing values on EUs may pose some problems, but no threat, when they occur infrequently.

     When each treatment is used in each block of the design, it is called a complete block design. It is not recommended to use more than 10 EUs in a block. If there are too many treatments it becomes necessary to use only a subset of them in each block. There are several classes of designs available for this purpose. Of these, the balanced incomplete block design has drawn much attention in the past 20 years and is considered optimum whenever it exists. This was used in Example 2.15. The study of the incomplete block designs is beyond the scope of this book.

Latin Square Design

Formally, a Latin square design is an arrangement of v treatments in a v row and v column square such that every treatment occurs once in each row and once in each column. For example, a Latin square design with $v = 5$ is

A	B	C	D	E
B	C	D	E	A
C	D	E	A	B
E	A	B	C	D
D	E	A	B	C

where the treatments are denoted by the letters A, B, C, D, and E. Latin square designs are applicable in experimental work in two contexts:

1. When the EUs have two different types of variations, one denoting the variation based on the rows and another on the columns, the treatments are assigned to the EU such that each of the row and column classifications receive each treatment exactly once. Thus, these designs are known in statistical literature as designs eliminating heterogeneity in two directions.

2. When only v EUs are available for experimental work on v treatments, and if it is possible to give each of the treatments to each EU sequentially, a Latin square design is used. As noted before the EUs in this context are called subjects.

Consider an experiment using five treatments and five subjects and consider using the design illustrated. Each row corresponds to a specified period of experimentation. On subject 1, treatments A, B, C, E, and D are given, respectively, in the five periods during the course of the experiment and data are recorded for each of the five periods. Because a treatment may produce a type of carry-over effect even after it is discontinued, enough washout time must be left between periods of application so that the subject reasonably returns to the original condition before receiving the next treatment. On subject 2, treatments B, C, D, A, and E are given, respectively, in the five periods. The treatment sequences given to the other subjects are determined by the columns 3, 4, and 5.

In experiments of this nature, Latin square designs are economical and are efficient in reducing the experimental variation and measuring the actual treatment differences more accurately. The randomization procedure for Latin square designs is more involved and will not be discussed here.

Crossover Design

The simplest design of this type is the two-treatment, two-period crossover. It is widely used in pharmaceutical trials and has been the target of severe criticisms in the industry. If A and B are two treatments, n subjects will be randomly divided into two groups of n_1 and n_2 subjects. Each member of the first group receives the treatment sequence A, B in that order for the two periods, while each member of the second group receives the treatment sequence B, A in that order for the two periods. One of the criticisms is that it is unethical to switch treatments in patients when a treatment is effective in the first period.

The second criticism is based on the interaction between periods and subjects and is beyond the scope of this book.

Examples showing the applications of these four designs will now be given.

Example 2.21. (Do Gentlemen Prefer Blondes?). Suppose a social scientist wants to determine the effect of women's hair color on the length of conversations men have with them. If one thinks that the nationality and age of men influence the conversational pattern, it is better to introduce them into the experiment to get the conclusions on a broader spectrum of men. Let the experimenter decide to use four nationalities, say English, French, German, and Italian. Also, let men from 4 different age groups, say 20–24, 25–29, 30–34 and 35–39 be considered for the experiment. The experiment will then have 16 volunteers, four men from each of the above mentioned nationalities and four from each of the above mentioned age groups. Furthermore, the four men from each nationality will belong to each of the age groups and the four men from each age group will belong to each of the nationalities. The four treatments can be considered as four women, one of each type: A = brunette, B = blonde, C = brown-haired, and D = red-haired. The response here is the time the conversation lasts without coming to a major pause. Because the EUs are classified based on two variables and the treatments should be used accounting for the two types of variations of EUs, the appropriate design for this experiment is a Latin square. One may conduct this experiment in the following manner:

Age group	Nationality			
	English	French	German	Italian
20–24	A	B	C	D
25–29	B	A	D	C
30–34	C	D	A	B
35–39	D	C	B	A

According to this plan, the brunette will be tested against the 20 to 24 year-old English, 25 to 29 year-old French, 30 to 34 year-old German,

and 35 to 39 year-old Italian men. The blonde will be used against the
20 to 24 year-old French, 25 to 29 year-old English, 30 to 34 year-old
Italian and 35 to 39 year-old German men. The brown haired girl will
be used against the 20 to 24 year-old German, 25 to 29 year-old
Italian, 30 to 34 year-old English and 35 to 39 year-old French men.
Finally, the red haired girl will be used against the 20 to 24 year-old
Italian, 25 to 29 year-old German, 30 to 34 year-old French and 35 to
39 year-old Englishmen. One may wonder whether this experiment
can be performed with four brunettes, four blondes, four brown haired,
and four red haired girls. The answer is no because it creates an extra
variation within each of the treatments.

Example 2.22. (Soft Drinks). Suppose one wants to know whether people like Coca-Cola, Pepsi-Cola, Seven-Up, ginger ale, or Sprite.
Let vending machines with five windows, with each machine showing
all five drinks, be randomly placed at several key locations of a city. At
the end of one month, the number of cans of each kind sold at each of
the locations is recorded. From these data, one can determine the most
sold drink after accounting for the random fluctuations. The design
used in this context is RBD. Each vending machine is a block and the
soft drinks are treatments. The experimenter is interested to compare
the sales to determine the most sold drink. To eliminate bias errors, the
investigator should keep a continuous supply of each drink at each
location and should not use more windows for one drink than for
others.

Example 2.23. (Pain Killers). Suppose one wants to know which
of the pain killers Bufferin, Excedrin, Empirin, and Tylenol give quick
relief for minor arthritic pains. One can choose a group of volunteers
who have arthritic pains and whose physical profiles are almost the
same. The volunteers will be divided into four, preferably, equal groups
and each volunteer of the first group is given Bufferin, each volunteer of
the second group is given Excedrin, each volunteer of the third group is
given Empirin and each volunteer of the last group is given Tylenol.
The response is the time, from the onset of pain until the relief is noted.
The design used here is CRD. If enough volunteers of similar health
profiles are not available, an RBD or more complicated design should
be used.

Example 2.24. (Animal Feeding Experiment). An animal nutritionist is interested in comparing two types of feed, A and B, in terms of the milk yield of cattle. A group of n animals, which are in similar health condition and that produce the same milk yield under normal conditions, will be randomly divided into two groups. Each animal in the first group will be fed with A in early lactation and will be switched to B in late lactation, whereas each animal in the second group is fed by B in early lactation and by A in late lactation. The average daily milk yield under each feed regimen for each animal will be recorded in the middle part of the feeding period. For example, if each feed is given for 3 months, the average milk yield data will be considered for the middle 1 month. By doing so one can eliminate the possible carry-over effect of the earlier feed into the latter feed. The design used here is a simple two treatment crossover design in two periods.

McLean and associates (1973) produced a check list of items, analogous to the ones discussed earlier for survey data, to be considered for planning an industrial experiment.

The nature of the EUs and the way the treatments are selected for the experiment affect the statistical analysis of the experiments. Some aspects of the analysis of this data will be discussed in Chapter 6.

Other methods of collecting data will be discussed in the subsequent sections.

PANEL STUDIES

Sometimes data will be continuously collected over a time period from a panel of experts. The same panel of experts will be used throughout the study. Such studies are also called longitudinal studies. A chocolate company may use a panel of tasters and each batch of chocolates coming from the production lines will be tested by the panel to check the quality of the taste. A similar design can be used to model the morale of employees in a company using a panel of employees who measure the morale on a ten-point scale over a period of time.

COHORT STUDIES

In the statistical sense a cohort is defined as a group of people within a geographically or other delineated population who experienced the

same significant life events within a given period of time. Cohort
studies refer to the studies involving groups of people over different
periods. The following illustrate the salient features of such studies.

Example 2.25.   (Abortions). Suppose a religious leader is interested in studying the abortion issue. Clearly, with changing times, the
magnitude of the problem changes and age of the women certainly also
has an influence on this. If the study is expected to be carried out over
a 20-year period from 1960 to 1979, data of the type indicated in Table
2.1 is needed. The data in the body of the table usually is taken from
published data either by census or sample survey. For example, one
may take the abortions/live birth ratio during 1960–1964 for women in
the age group 15–19 throughout the country or from a given hospital.
When data are collected by a sample survey, it is not necessary that
they are collected form the same sample SUs and in this respect, cohort
studies are different from longitudinal studies. The variations in the
ratios in Table 2.1 are produced because of three influences: (1) age
effects, (2) period or time effects, and (3) effects because of cohorts,
or cohort effects. The age effects for a given period can be studied by
examining the data through the columns. The period effects for given
age groups can be examined through the rows. The diagonal entries
from the top left hand corner to the bottom right hand corner indicate
the cohort effects. The women in 15–19 year age group in 1960–1964
are precisely the women in 20–24 year age group in 1965–1969 and

Table 2.1   Ratio of Abortions to Live Births

Age of Women	Year			
	1960–1964	1965–1969	1970–1974	1975–1979
15–19				
20–24				
25–29				
30–34				
35–39				

they are the ones in 25–29 year age group in 1970–1974 and finally, in the 30–34 year age group in 1975–1979.

PUBLISHED DATA SOURCES

Investigators who do not collect their own data by surveys and experiments must depend upon available outside data sources. Already established data bases are needed in several studies. It is physically impracticable to complete a cohort study without using an existing data base.

Most of the American national survey data is available from two major survey data archives: the Roper Public Opinion Research Center in Williamstown, Massachusetts, and the Survey Data Archive of the Institute for Social Research at the University of Michigan, Ann Arbor.

The Gallup Poll data is published in a monthly periodical entitled *The Gallup Opinion Index* from 1966. Other useful survey data are held by the National Opinion Research Center in Chicago, the Louis Harris polling organization in New York and the Bureau of the Census in Washington.

The *World Almanac, U.S. Meteorological Yearbook, County and City Data Book of the U.S. Bureau of the Census, Vital Statistics of the United States, The Sports Encyclopedia of Baseball, The Wall Street Journal, Statistical Year Book of the U.N. Statistical Office, Statistical Abstract of the United States of U.S. Bureau of the Census, Demographic Year Book of the United Nations Statistical Office, Commodity Year Book of the Commodity Research Bureau,* and *A Statistical History of the American Presidential Elections* are some publications where the investigator may get some useful published data.

REFERENCES

Cochran, W.G. and Cox, G.M. (1957). *Experimental Designs*. Wiley, New York.

Ford, B.L. and Tortora, R.D. (1978). A consulting aid to sample design. *Biometrics* 34. 299–304.

McLean, R.A., Anderson, V.L., Bayer, H.S., and McElrath, G.W. (1973). A scientific approach to experimentation for consulting statisticians. *J. Quality Technol.* 5. 1–6.

Raghavarao, D. (1971). *Constructions and Combinatorial Problems in Design of Experiments*. Wiley, New York.

Smith, L.L., Federer, W.T., and Raghavarao, D. (1976). A comparison of three techniques for eliciting truthful answers to sensitive questions. *Proc. ASA Soc. Statist. Sect.*, 447–452.

Warner, S.L. (1965). Randomized response. A survey technique for eliciting evasive answer bias. *J. Am. Statist. Assoc.* 60. 63–69.

APPENDIX

Number of Replications for an Experimental Design

To decide the exact number of replications in a design, one should have a knowledge of the level of significance and power of the test. An interested, statistically mature reader may consult Cochran and Cox (1957, pp. 17–29) for this purpose. A heuristic method of determining the replication numbers will now be discussed. In Chapters 4 and 6 the reader will be introduced to certain statistical testing procedures. The tests are based on using published tables, where the table entries show a quantity known as degrees of freedom. The table values fluctuate widely when the degrees of freedom are small and more or less stabilize when the degrees of freedom are at least 12. With this in mind, the replicates of a treatment in each design can be determined. With v treatments in a CRD using equal number of replications, r, for each treatment, one should use

$$r \geq \frac{12}{v} + 1 \qquad (2A.1)$$

With v treatments in a RBD using r replications for each treatment, the number of blocks is also r and is

$$r \geq \frac{12}{v-1} + 1 \qquad (2A.2)$$

In a Latin square design using v treatments, the experimenter has no choice for the number of replications and v replications are required. The number of subjects, n, in a two-period crossover design is the same as the number of replications of the treatments and should be taken as

$$n \geq 13 \qquad (2A.6)$$

The number of replications given is only a guideline for the experimenter.

EXERCISES

In Exercises 1–5 indicate the appropriate sampling method: (a) Simple Random Sampling, (b) Systematic Sampling, (c) Stratified Sampling, and (d) Cluster Sampling.

1. To estimate the proportion of drivers wearing shoulder belts on an interstate highway
2. To estimate the average life of a given brand of 100 watt electric bulbs
3. To estimate the average number of study hours per week by college students
4. To estimate the average annual income of residents in a city
5. To estimate the number of children who will go to grade 1 next year in a given community

In Exercises 6–10 indicate the most appropriate experimental design: (a) CRD, (b) RBD, (c) Latin Square Design and (d) Crossover Design.

6. To compare four gas additives in increasing octane measure when four cars of different makes are available for the test
7. To compare the effectiveness of three teaching methods in a college to which students with a 1200 or above SAT score (or its equivalent) are admitted
8. To compare the effect of sales by packing potatoes in 5 lb bags, 10 lb bags and, 25 lb bags based on sales in 10 stores.
9. To compare two laboratory techniques for making WBC counts based on 20 blood samples
10. To compare the effect of two drugs on anginal attacks based on 20 patients suffering from heart problems

Answers

1. (b); 2. (a); 3. (c); 4. (c); 5. (a); 6. (c); 7. (a); 8. (b); 9. (b); 10. (d).

3

Summarization of Data

After collecting data, the investigator prepares appropriate data files, and forms tables and charts showing the patterns of the observations. In addition to the tables and charts, one also forms some values, based on the observations, that indicate a general idea of the underlying phenomenon represented by the data. These aspects will be discussed in this chapter.

FREQUENCY TABLES, LINE DIAGRAMS, AND HISTOGRAMS

Frequency tables, line diagrams, and histograms are widely used to present a summary view of statistical data. When the observations are qualitative, the number of observations (called frequency) in each of the qualitative categories are presented and line diagrams are used to graphically represent the data. An illustration of this situation is discussed in Example 3.1:

Example 3.1.   (Democratic votes in 1980 and 1984 from counties in Massachusetts). Consider Table 3.1.

The data are the number for voters in each county in 1980 and in 1984. The tabulation is made showing the county categories and indicating the number of voters in each county. The numbers shown in columns 3 and 4 are called the frequencies. One easily notes that the total number of voters in 1984 is more than in 1980, and, hence, one hopes to observe more Democratic votes in 1984 in each county than in 1980 if there is no crossover of voting pattern or people changing their county of residence. In Brockton county there were 12,751 votes in

Table 3.1   Votes for Democratic Presidential Candidate in 1980 and 1984 in Selected Counties in Massachusetts and Totals for the State

		Votes in	
Number	County	1980	1984
1	Boston	95,133	131,015
2	Brockton	12,751	13,983
3	Cambridge	24,337	32,409
4	Fall River	19,644	20,898
5	Framingham	12,275	14,368
6	Lawrence	12,145	10,986
7	Lowell	16,353	15,033
8	Lynn	15,777	17,099
9	New Bedford	18,014	22,067
10	Newton	20,173	27,299
11	Quincy	17,977	19,275
12	Somerville	16,931	21,056
13	Springfield	26,414	29,379
14	Worcester	31,146	31,100
	State totals	1,053,802	1,219,513

Source: The World Almanac and Book of Facts, 1985, p. 52. Reproduced with the permission of the publisher.

1980 and 13,983 votes in 1984. Although there is an apparent increase
of votes in 1984 in this county, it may not indicate a real increase
because the total voters have increased in 1984. Similarly the real
decrease of voters in 1984 in Lowell county is much more than the
obvious decrease of $16,353 - 15,033 = 1,320$. To compare the 1980
and 1984 votes in these counties, it is essential that the votes should be
converted to percentages. Surprisingly, from Table 3.2 one notes that
in Brockton county the percentage of votes went down in 1984 compared to 1980. A graphic presentation of the data of Table 3.2 is a line
diagram as shown in Figure 3.1. In this setting one may want to
graphically visualize the differences in percentage of votes and this can
be achieved from Figure 3.2 in which the line above the X-axis at a

Table 3.2 Percentage of Votes for Democratic
Presidential Candidates in 1980 and 1984 in Selected
Counties of Massachusetts

Number	County	Percentage votes in 1980	1984	Difference 1980–1984
1	Boston	9.03	10.74	−1.71
2	Brockton	1.21	1.15	0.06
3	Cambridge	2.31	2.66	−0.35
4	Fall River	1.86	1.71	0.15
5	Framingham	1.16	1.18	−0.02
6	Lawrence	1.15	0.90	0.25
7	Lowell	1.55	1.23	0.32
8	Lynn	1.50	1.40	0.10
9	New Bedford	1.71	1.81	−0.10
10	Newton	1.91	2.24	−0.33
11	Quincy	1.71	1.58	0.13
12	Somerville	1.61	1.73	−0.12
13	Springfield	2.51	2.41	0.10
14	Worcester	2.96	2.55	0.41

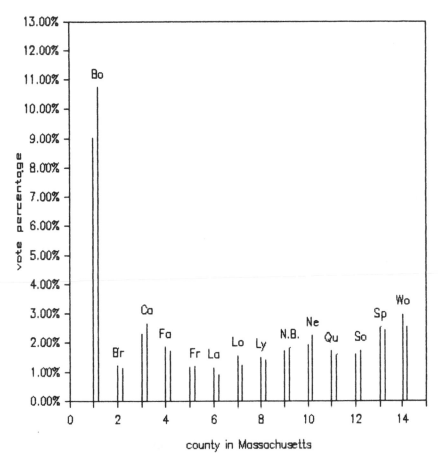

Figure 3.1   Percentage of Democratic vote in 1980 compared to 1984.

county indicates a decrease in the percentage of 1984 votes over the percentage of 1980 votes, and a line below the X-axis at a county indicates an increase in the percentage of 1984 votes over that of 1980 votes.

With quantitative data, there are no natural classes in which the frequencies are counted. The investigator arbitrarily and conveniently makes up the classes and counts the frequency and percentage of observations in each class. The terminology and the concepts used in this context are explained in Examples 3.2 and 3.3.

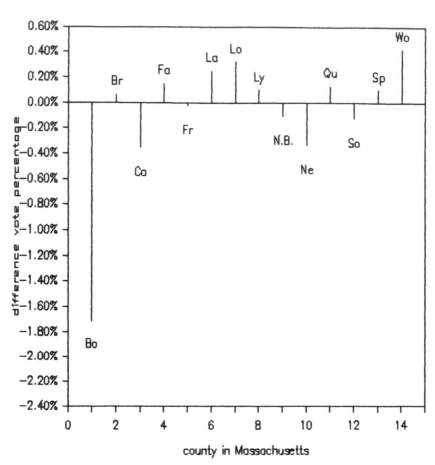

Figure 3.2   Percentage of difference in Democratic vote in 1980 compared to 1984.

Example 3.2.   (Drainage Areas for 30 of the Largest Rivers in the US). The drainage area for 30 of the largest US rivers is summarized in Table 3.3. The drainage area is divided into eight distinct classes. In the first class 10,000 and 20,000 are, respectively called the lower and upper class boundaries and their difference 20,000 − 10,000 = 10,000 is called the class length of the first class. Three rivers have drainage area between 10,000 and 20,000. Although the exact drainage areas of these three rivers are known in the raw data (that is, untabulated data), that information is masked in the frequency Table 3.3. The drainage areas of the three rivers are assumed to be equally spread over the entire class length of 10,000 to 20,000. Although 10,000 is shown as the lowest class boundary of the very first class, it is not necessary that some river should have a drainage area of 10,000. The 10.0 in the third column indicates that 10% of the 30 largest US rivers have drainage areas between 10,000 and 20,000. The second class is also of class length 30,000 − 20,000 = 10,000 and has 20,000 as the lower class boundary and 30,000 as the upper class boundary. Ten rivers have drainage areas between 20,000 and 30,000, and the actual drainage areas of these 10 rivers are assumed to be equally spread out in that class. The third column indicates that 33.3% of the 30 largest US rivers have drainage areas between 20,000 and 30,000. A similar interpretation can be given to each of the remaining six classes. A graphic representation of the data is a histogram given in Figure 3.3, in which the areas of the rectangles corresponding to the classes, are proportional to the class frequencies.

Figure 3.3 does not give a good visual portrayal of the data. The histogram is squashed together in the left part of the graph. One may show drainage area up to 0.5 million acres and get a somewhat better

Table 3.3 Drainage Area (in Acres) Under 30 Largest U.S. Rivers

Drainage Area	Frequency	Percentage	Cumulative frequency
10,000–20,000	3	10.0	3
20,000–30,000	10	33.3	13
30,000–40,000	2	6.7	15
40,000–50,000	5	16.7	20
50,000–100,000	2	6.7	22
100,000–250,000	3	10.0	25
250,000–500,000	3	10.0	28
500,000–1,500,000	2	6.7	30
Total	30	100.1[a]	

Source: The World Almanac and Book of Facts, 1985, p. 628.
[a]Rounding error.

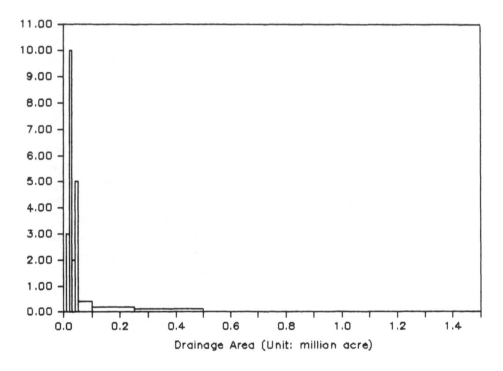

Figure 3.3   Drainage area under 30 largest US rivers.

looking histogram; but it does not indicate that some rivers have the potential of giving drainage areas of more than 0.5 million acres. Data of this type with large spread of observations with small frequencies in classes pose a problem for graphic representation. Figure 3.4 gives a good visual presentation of a histogram and will be discussed in the next example.

The last column of Table 3.3 gives the cumulative frequencies, that is, the total frequency below the upper class boundary of the class. There are three rivers with drainage areas less than 20,000; thirteen rivers with drainage areas less than 30,000; fifteen rivers with drainage areas less than 40,000; twenty rivers with drainage areas less than 50,000; twenty two rivers with drainage areas less than 100,000; twenty five rivers with drainage areas less than 250,000; twenty eight rivers with drainage areas less than 500,000; and all the thirty rivers with drainage areas less than

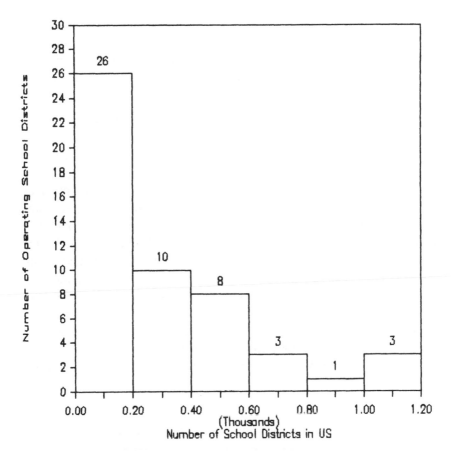

Figure 3.4 Frequency histogram.

1,500,000. Also note that even though 1,500,000 is the upper class boundary of the last class, it is not necessary that at least one river should have a drainage area of 1,500,000.

In Table 3.3, the class lengths are unequal and, in general, in a frequency table the class lengths may or may not be equal. It is also possible to find open-ended classes at the beginning or at the end of the table. The first class of Table 3.3 may be marked as ''<20,000'' and the last class as ''>500,000.'' It is recommended that the boundaries of the classes be kept distinct from the observations. If the boundaries are taken to one more decimal of accuracy than the observations, no data falls on the boundaries. Ignoring this rule, if boundaries are chosen so that an observation turns out to be one of the boundary values, then the observation is counted in the class in which it is the lower class boundary. General guidelines for forming frequency tables and histograms will be given in the chapter Appendix.

Example 3.3. (Operating Local School Districts in US). Table 3.4 gives a frequency table of the number of operating school districts in the 50 states plus the District of Columbia. There are twenty six states with the number of school districts between 0 and 200; ten states with the number of school districts between 200 and 400; eight states with the number of school districts between 400 and 600; three states with the number of school districts between 600 and 800; one state with the number of school districts between 800 and 1000; and three states with the number of school districts between 1000 and 1200. Alternatively, one may think that 51% of the states have the number of school districts between 0 and 200; 19.6% have the number of school districts between 200 and 400; 15.7% have the number of school districts between 400 and 600; 5.9% have the number of school districts between 600 and 800; 2.0% have the number of school districts between 800 and 1000; and 5.9% have the number of school districts

Table 3.4 Number of Operating School Districts in U.S.

Number of school districts	Frequency	Percentage	Cumulative frequency
0– 200	26	51.0	26
200– 400	10	19.6	36
400– 600	8	15.7	44
600– 800	3	5.9	47
800–1,000	1	2.0	48
1,000–1,200	3	5.9	51
Total	51	100.1[a]	

Source: *The World Almanac and Book of Facts, 1985*, p. 244.
[a]Rounding error.

between 1000 and 1200. It is not necessary that a state has 0 school districts because 0 is the lowest class boundary of the beginning class. It is also not necessary that a state has 1200 school districts because 1200 is the upper class boundary of the last class. The class lengths of all classes are equally spaced and the frequency in each class is assumed to be evenly spread out over the class lengths. Finally, twenty six states have less than 200 school districts; thirty six states have less than 400 school districts; forty four states have less than 600 school districts; forty seven states have less than 800 school districts; forty eight states have less than 1000 school districts; and all states have less than 1200 school districts. A histogram for the data of Table 3.4 is Figure 3.4.

A more informative tabulation of data other than the frequency table, is the stem-and-leaf plot now discussed.

STEM-AND-LEAF DISPLAYS

Stem-and-leaf plots are convenient summary presentations of the data made by quickly jotting down the observations in a convenient and understandable way. It is similar to a frequency table, but uses the actual data to create the display. The simplest forms of the displays are considered in this section and the reader is referred to Tukey (1977) for other variations of these presentations.

Example 3.2 (*continued*).   A stem-and-leaf display of the drainage area of the 30 largest rivers in the US is given in Table 3.5. The digits to the left of the line of Table 3.5 are stems and each digit to the right of the line is a leaf. The table will be read by combining each number to the left of the vertical line with each number to the right of the vertical line in the indicated units. In this case one can interpret the table by reading the stems in 10s and leaves as 1s in units of 10,000. For example, the second line of the above table indicates that 10, 11, and 16 ten thousand acres are drainage areas of three rivers.

Example 3.3 (*continued*).   The stem-and-leaf display of the number of local operating school districts of the 50 states plus the District of Columbia is given in Table 3.6. By looking through the fourth stem, one notes that six states have 300, 304, 306, 309, 351, and 371 operating school districts. The leaves on the stem marked 6 indicate

Table 3.5 Stem-and-Leaf Display of the Drainage Area (in Rounded Units of 10,000 Acres) of the 30 Largest US Rivers

0	1,2,2,2,2,2,2,2,3,3,3,3,3,3,3,4,4,4,5,5,9
1	0,1,6
2	0,6
3	0,3
4	
5	3
6	
7	
8	
9	
10	
11	
12	5

Source: *The World Almanac and Book of Facts, 1985*, p. 628.

Table 3.6 Stem-and-Leaf Display of the Number of Operating School Districts in 50 States Plus the District of Columbia

0	01,01,17,19,24,40,40,49,53,55,66,67,89,92
1	15,28,35,43,46,53,58,65,80,81,87,87
2	08,33,47,92
3	00,04,06,09,51,71
4	33,35,41
5	00,48,61,72,82
6	15,19
7	15
8	
9	68
10	08,33,75

Source: *The World Almanac and Book of Facts, 1985*, p. 244.

that two states have 615 and 619 number of local operating school districts. The leaves on other stems can be similarly interpreted.

In Tables 3.5 and 3.6 only one line is used for each stem. It is possible to use more than one line for each stem. The Minitab computer program discussed at the end of this chapter uses 1, 2 or, 5 lines per stem depending on the number of values present.

So far the data has been expressed for only one variable. Often, one may need to form summary tables representing the data on two or more variables and such tables in the literature are known as contingency tables.

TWO-WAY AND HIGHER-WAY CONTINGENCY TABLES

When each unit has observations based on two or more variables, it is customary to represent the frequency of observations that belong to specified classes of the variables and such tables are called contingency tables. The following examples illustrate this type of tables.

Example 3.4. (Average Life Span and Gestation Period of Selected Animals). Table 3.7 gives the number of animals with given average life span (in years) and gestation period (in days). From Table 3.7 one notes that 20 out of 41 animals have an average life span of 0–15 years with a gestation period of 0–250 days; 9 have an average life of

Table 3.7 Contingency Table of Average Life (yrs) and Gestation Period (days) of 41 Different Animals

Gestation Period (days)	Average life (yrs)			
	0–15	15–30	30–45	Totals
0–250	20	9	0	29
250–500	4	7	0	11
500–750	0	0	1	1
Totals	24	16	1	41

Source: The World Almanac and Book of Facts, 1985, p. 186.

15–30 years, gestation period 0–250 days; 4 have an average life of 0–15 years, gestation period 250–500 days; 7 have an average life of 15–30 years, a gestation period of 250–500 days; and 1 has an average life of 30–45 years, gestation period of 500–750 days. The row and column totals given in the table are called marginal totals; they indicate the frequency associated with the corresponding classifications. Twenty-four animals have an average life of 0–15 years, 16 animals have an average life of 15–30 years; and one animal has an average life of 30–45 years. Further, 29 animals have a gestation period of 0–250 days; 11 animals have a gestation period of 250–500 days; and one animal has a gestation period of 500–750 days. Furthermore, it appears that animals with shorter life spans have a shorter gestation period and those with longer life spans have longer gestation periods, although such observations need support from statistical analysis using methods discussed later in this book (see Review Exercises 56 and 57).

Example 3.5. (Childhood and Adulthood Diets). Suppose a nutritionist is interested in studying the effect of the childhood diet (e.g., vegetarian or nonvegetarian) on the preferred adult diet. Data will be collected on sampled individuals regarding their childhood diet and their preferred diet. The results can be summarized in Table 3.8. Table 3.8 indicates that of the 60 people interviewed, nine prefer a vegetarian diet and 51 prefer a nonvegetarian diet. Of them, 10 people received a vegetarian diet and fifty received a nonvegetarian diet when they were children. Of the 10 people who received a vegetarian diet in their childhood, 4 people still prefer a vegetarian diet, while 6 prefer a

Table 3.8 Artificial Data on Childhood and Preferential Adulthood Diets

Childhood Diet	Adult preferred diet		Totals
	Vegetarian	Nonvegetarian	
Vegetarian	4	6	10
Nonvegetarian	5	45	50
Totals	9	51	60

nonvegetarian diet. Of the 50 people who received a nonvegetarian diet, 45 still prefer a nonvegetarian diet and 5 prefer a vegetarian diet.

Example 3.6. (Instructors and Grades). Usually, in most universities, several sections of the basic courses will be offered and different instructors will be assigned to teach each section. In such situations the administration likes to examine the grading pattern and set uniform standards. The data of the grades assigned by instructors in a semester can be arranged in a contingency table. Table 3.9 is one such illustration. The number of students who got A, B, C, D, F, and W grades are 23, 14, 17, 5, 11, and 25 respectively. Twenty students were taught by Instructor I, 40 students were taught by Instructor II, and 35 students were taught by Instructor III. Of the 23 A grades, Instructor I gave 15; II gave 2; and III gave 6. A similar interpretation can be given for the other numbers in the table. Whereas it is expected that more students get C grades in such courses with fewer students getting A and F grades, the overall performance in the whole course is just the opposite. If the sections contain a similar mix of students, the performance of students in the sections must be similar; it then seems that Instructor I gives too many high grades and Instructor II gives too many failing grades. More insight into the grading pattern can be obtained by performing the necessary statistical analysis (see Review Exercise 58).

Sometimes the frequencies have to be summarized using three or more variables. They are known as higher–way contingency tables.

Table 3.9 Artificial Data of the Grades Given By Three Instructors

Instructor	Grades						Totals
	A	B	C	D	F	W	
I	15	4	1	0	0	0	20
II	2	3	4	1	10	20	40
III	6	7	12	4	1	5	35
Total	23	14	17	5	11	25	95

Though the statistical analysis of higher–way tables are not discussed in this text, an illustration of a three-way table is given in the next example.

Example 3.7. (Infants Survival and Parental Care at Two Clinics). Death or survival of infants at two clinics and the parental care received by them is summarized in Table 3.10. From Table 3.10 it is clear that the study involved seven hundred fifteen infants: 476 infants were considered from clinic A and 239 from clinic B. Of the 715 infants, 26 infants died and 689 survived. Of the 26 dead infants, 7 were from clinic A and 19 from clinic B. Of the 7 dead infants at clinic A, 3 received less parental care and 4 received more parental care. Of the 19 dead infants at clinic B, 17 received less parental care and 2 received more parental care. Of the 689 surviving infants, 469 were from clinic A and 220 from clinic B. Of the 469 surviving infants at clinic A, 176 received less parental care and 293 received more parental care. Of the 220 surviving infants at clinic B, 197 received less parental care and 23 received more parental care. Of the total 476 infants at clinic A, 179 received less parental care and 297 received more parental care. Of the total 239 infants at clinic B, 214 received less parental care and 25 infants received more parental care. In Table 3.10 each infant is classified based on three variables: 1. Death or survival, 2. Clinic, and 3.

Table 3.10   Three-Way Contingency Table on Infants Survival and Parental Care at Two Clinics

Infants' survival	A Parental care less	A Parental care more	A Subtotal	B Parental care less	B Parental care more	B Subtotal	Totals
Died	3	4	7	17	2	19	26
Survived	176	293	469	197	23	220	689
Totals	179	297	476	214	25	239	715

Source: Bishop, Fienberg, and Holland (1975, p. 41) from Bishop (1969). Reproduced with the permission of the authors and publisher.

Parental care. Thus, the table is known as a three-way contingency table. It is also called a three-dimensional array. From Table 3.10 one can form three two-way contingency tables as indicated in Tables 3.11A, 3.11B, and 3.11C. By looking at Table 3.10, one may er-

Table 3.11A Infants' Survival and Parental Care

Parental care	Infants' survival		Totals
	Died	Survived	
Less	20	6	26
More	373	316	689
Totals	393	322	715

Table 3.11B Infants' Survival and Clinics

Clinics	Infants' survival		Totals
	Died	Survived	
A	7	469	476
B	19	220	239
Totals	26	689	715

Table 3.11C Parental Care and Clinics

Clinics	Parental care		Totals
	Less	More	
A	179	297	476
B	214	25	239
Totals	393	322	715

roneously think that survival rate depends on parental care, because twenty deaths of twenty six total deaths, occurred in connection with less parental care. Bishop, Fienberg and Holland (1975) statistically concluded that it is not so.

TWO-AND HIGHER-WAY SUMMARY TABLES

When experiments are done using several levels of different factors, the mean responses corresponding to the combinations of the levels of different factors must be summarized in appropriate tabular form. Such presentations are discussed in the following examples.

Example 3.8.   (Ascorbic Acid Levels). The ascorbic acid level was determined (mg/100 g) in snap beans stored at three temperatures and over four time periods. The results of such an experiment can be summarized as in Table 3.12. From Table 3.12 one understands that after 2 weeks of storage, the ascorbic acid content is 15.0 if stored at 0°F; 15.0 at 10°F; and 11.3 at 20°F. After 4 weeks of storage, it is 15.7 if stored at 0°F, 14.3 at 10°F, and 9.3 at 20°F. After 6 weeks of storage, it is 15.3 at 0°F, 13.7 at 10°F, and 7.0 at 20°F. After 8 weeks of storage, it is 15.3 at 0°F, 12.3 at 10°F, and 5.3 at 20°F. On the whole, it is 15.3 at 0°F, 13.8 at 10°F, and 8.2 at 20°F. Further, on the

Table 3.12   Average Ascorbic Acid Determination (mg/100 g) in Snap Beans at Three Temperatures and Four Periods

Temperature (°F)	Weeks of storage				
	2	4	6	8	Mean
0	15.0	15.7	15.3	15.3	15.3
10	15.0	14.3	13.7	12.3	13.8
20	11.3	9.3	7.0	5.3	8.2
Mean	13.8	13.1	12.0	11.0	12.5

Source: Snedecor and Cochran (1980, p. 311) from Snedecor (1947). Reproduced with the permission of the publisher.

whole, it is 13.8 after 2 weeks of storage, 13.1 after 4 weeks, 12.0 after 6 weeks, and 11.0 after 8 weeks. The individual observations giving the cell means as indicated (or the cell means) can be statistically analyzed by the methods outlined in Chapter 6 to test the effects of length of storage and temperature on ascorbic acid content in snap beans (see Review Exercise 59).

Example 3.9. (Effect of Lysine, Methionine and Protein on Weight Gain in Pigs). The gain in weight of male pigs' was determined using supplementary lysine, methionine and protein in their diet, and the results are summarized in Table 3.13. Any entry in the table is the average weight gain of pigs by feeding corresponding levels of methionine, protein, and lysine. For example, by giving 0.25% methionine, 12% protein and 0.10% lysine as a supplementary diet, the average gain in weight is 1.38. By giving 0% methionine, 14% protein and 0.05% lysine as a supplementary diet, the average gain in weight is 1.54.

In the next two sections the reader will be introduced to numerically summarizing the data.

Table 3.13   Average Daily Weight Gains of Pigs Fed Supplementary Lysine, Methionine and Protein

Lysine (%)	Methionine (%)					
	0		0.25		0.50	
	Protein (%)		Protein (%)		Protein (%)	
	12	14	12	14	12	14
0	1.04	1.49	1.04	1.25	1.03	1.46
0.05	1.15	1.54	1.12	1.29	1.04	1.52
0.10	1.18	1.23	1.38	1.31	1.27	1.43
0.15	1.11	1.05	1.26	1.41	1.25	1.45

Source: Snedecor and Cochran (1980, p. 319) from Iowa Agricultural Experiment Station, 1952. *Anim. Husb. Swine Nutr. Exp.* 577. Reproduced with the permission of the publisher.

CENTRAL TENDENCY MEASURES

The measures of central tendency indicate the location of data. When test papers are returned, the students almost always ask "What is the average score?" If the teacher informs them 70, most students presume that they might have gotten a score of about 70. Some students might have gotten scores much higher than 70 and some much lower than 70. The high scores compensate the low scores and the average score is 70, and it is a typical value characterizing the test score data. The average value of a data balances the histogram. Students who have taken the Scholastic Aptitude Test (SAT) get their percentile score. If a student is in the 90th percentile, it means that nearly 90% of the students taking the test got a score less than that individual and nearly 10% of the students taking the test got a score more than that individual. A shirt manufacturer makes shirts with different neck sizes based on the demand. If a 15½ inch neck size is required by more people, he manufactures more 15½ inch neck size shirts and the mode of the neck size is 15½ inches. A Legislation to help the 50% lower-income families uses the median family income as a base. If 15,000 dollars is the median income of families in a community, nearly 50% of the families in that community earn less than 15,000 dollars while nearly 50% of the families earn more than 15,000 dollars. These ideas will be discussed extensively in this section.

The main purpose of the measures of central tendency is to summarize the data by a typical value. The commonly used measures are: *mean, median, mode, quartiles,* and *percentiles.*

The *mean* is simply the arithmetic average of the data. It makes use of all of the observations; however, is easy to see that a single extreme observation (whether very large or very small) considerably affects the mean. When the observations are more or less similar, the mean is an excellent measure of central tendency.

The *median* is the value such that nearly 50% of the observations are below it and nearly 50% of the observations are above it. It does not take the actual observations into consideration. It is calculated after ranking the observations and is based on the relative ranked positions. It is not affected by extreme values and, in this respect, it is a more stable measure than the mean. Graphically, if we consider the area of the bars in a histogram, the areas below the median and above the median are equal. When the data have extreme values, the median is a recommended measure of central tendency.

The calculation of mean and median for ungrouped data, and mean for grouped data will be discussed in the Appendix.

The *mode* is that observation that occurs more frequently in any data set. In case of ties, that is, when several observations each of which occur equally often and more than other observations, all such observations are considered as modes. A data set may not even have a mode, or have one mode or have several modes. A data set with 1 mode is called unimodal. In a histogram, the mode is a relatively peak point, that is, it has the largest frequency in its immediate neighborhood. The mode can be changed by changing the starting boundary and class length in a histogram. It is a very unstable measure of central tendency for a data set.

A *data set* has three quartiles Q_1, Q_2, and Q_3; nearly 25% observations are below Q_1, nearly 25% are between Q_1 and Q_2, nearly 25% are between Q_2 and Q_3, and nearly 25% observations are above Q_3. Clearly Q_2 is also the median for the data.

A data set has 99 percentiles p_1, p_2, . . . , p_{99}. Between any two successive percentile values there are nearly 1% of the observations. For example, nearly 1% of the observations are between p_{58} and p_{59}.

A graphic representation of these summary values can be given by box-and-whisker plots. In these plots the smallest observation, largest observation and the three quartiles are presented in a box form with two whiskers.

The ideas presented thus far will now be illustrated through examples.

Example 3.10. (Per Capita Income in the 50 States and the District of Columbia). From *The World Almanac and Book of Facts, 1985,* p. 162, the per capita income (in dollars) in 1983 of the 50 states plus the District of Columbia are

14,895; 9,847; 13,264; 12,021; 11,670; 9,979; 12,665; 15,744; 12,994; 14,122; 12,990; 11,488; 12,405; 10,476; 11,466; 11,216; 11,352; 10,705; 12,247; 11,913; 10,969; 11,212; 11,666; 9,847; 9,242; 8,967; 11,593; 10,379; 9,397; 10,270; 8,098; 9,787; 9,187; 9,549; 12,116; 9,159; 10,656; 9,640; 10,963; 11,685; 12,770; 9,555; 9,949; 8,993; 11,911; 13,257; 12,451; 10,740; 12,177; 17,194; 12,114.

For this data

$$\text{Mean} = \$11,352; \text{Median} = \$11,352; \text{Mode} = \$9,847,$$
$$Q_1 = \$9,847; Q_2 = \$11,352; Q_3 = \$12,247.$$

Excepting mode the remaining measures can be obtained by using the Minitab computer program as described in the Appendix. (For fuller details see Ryan, Joiner, and Ryan (1985)). Mode is that observation which occurs more times than other observations and can be obtained by examining the data. The mean per capita income of the states is $11,352. Some states have more than $11,352 and some have less than $11,352 per capita income and they balance out. Twenty-five states (nearly 50% of the states) have a per capita income less than $11,352

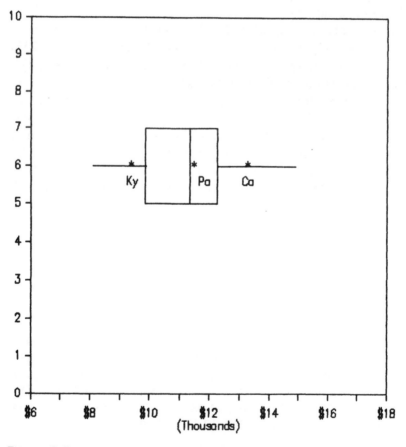

Figure 3.5   Box-and-whisker plot for per capita income.

and twenty five states (nearly 50% of the states) have a per capita income of more than $11,352. Most states have a per capita income $9,847. In this data only two states have $9,847 per capita income while no other per capita income observation is shared by two states. Nearly 25% of the states have a per capita income less than $9,847, nearly 25% of the states have a per capita income between $9,847 and $11,352, nearly 25% of the states have a per capita income between $11,352 and $12,247, and nearly 25% of the states have a per capita income more than $12,247. A schematic plot of per capita incomes of the 50 states plus the District of Columbia is given in Figure 3.5.

In Figure 3.5 the rectangular box runs from Q_1 to Q_3. The line within the box denotes the median, Q_2. Whiskers run out from this box to the adjacent values, which are the smallest and largest observations of the data. Pennsylvania, with a per capita income of $11,488, belongs to the third quarter of the richest per capita income states. California, with a per capita income of $13,257, is in the top 25% of the richest states in U.S. Kentucky, with a per capita income of $9,397, is in the lowest 25% of the poorest states in U.S.

Example 3.11. (Per Capita Income in Different Regions of US). Figure 3.6 provide a schematic representation of box-and-whisker plots in the eight regions. Each of the states are also marked on the graph with the following key:

1.	Alabama	18.	Kentucky	35.	North Dakota
2.	Alaska	19.	Louisiana	36.	Ohio
3.	Arizona	20.	Maine	37.	Oklahoma
4.	Arkansas	21.	Maryland	38.	Oregon
5.	California	22.	Massachusetts	39.	Pennsylvania
6.	Colorado	23.	Michigan	40.	Rhode Island
7.	Connecticut	24.	Minnesota	41.	South Carolina
8.	Delaware	25.	Mississippi	42.	South Dakota
9.	District of Columbia	26.	Missouri	43.	Tennessee
10.	Florida	27.	Montana	44.	Texas
11.	Georgia	28.	Nebraska	45.	Utah
12.	Hawaii	29.	Nevada	46.	Vermont
13.	Idaho	30.	New Hampshire	47.	Virginia
14.	Illinois	31.	New Jersey	48.	West Virginia
15.	Indiana	32.	New Mexico	49.	Washington
16.	Iowa	33.	New York	50.	Wisconsin
17.	Kansas	34.	North Carolina	51.	Wyoming

From Figure 3.6 one can visualize the relative standing of each state in its own region as well as in other regions.

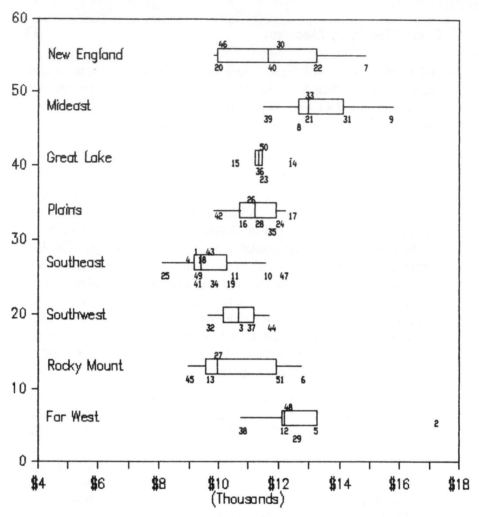

Figure 3.6 Box-and-whisker plots for per capita income in different US regions.

Example 3.12. (Temperatures at Selected US Cities). From *The World Almanac and Book of Facts, 1985*, p. 758, the normal monthly temperatures (in °F) for the twelve months January through December in Philadelphia are

$$32, 34, 42, 53, 63, 72, 77, 75, 68, 57, 46, 35$$

For this data

$$\text{mean} = 54.5 \qquad \text{median} = 55$$
$$Q_1 = 36.75 \qquad Q_2 = 55 \qquad Q_3 = 71$$

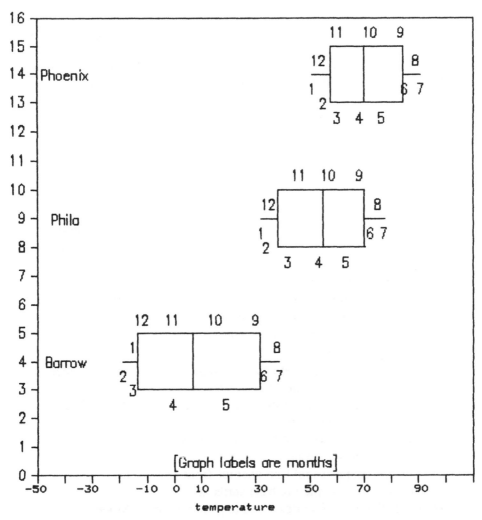

Figure 3.7 Box-and-whisker plot of temperatures at Barrow, Philadelphia, Phoenix.

Some months are hotter than 54.5°F and some are colder than 54.5°F, and they balance out. For 6 months, the temperatures are below 55°F, and for other 6 months they are above this temperature. For 3 months the temperatures are less than 36.75°F, for three months they are between 36.75°F and 55°F, for three months they are between 55°F and 71°F, and for three months they are higher than 71°F. The normal monthly temperatures can be schematically represented by box-and-whisker plots. In Figure 3.7, the temperatures of Barrow, Alaska,

Philadelphia, Pennsylvania, and Phoenix, Arizona are shown. In that display the months January through December are shown by numbers 1 through 12.

The appropriateness of a given central tendency measure in a specific problem must be carefully examined by the investigator. Otherwise, there is a danger of incorrectly interpreting the data. In the following sets of different scenarios, suggested central tendency measures are given:

Problem	Central Tendency Measure
A refrigerator company wants to decide the capacity of refrigerators with greatest demand	Mode
A transit company wants to run another bus if necessary demand exists	Median
Two sections of students are to be compared on the basis of their scholastic performance	Mean or median
To report the performance of students at a National level examination	Percentiles
To determine the income of a motel based on the number of daily patrons	Mean
25% of the people should be exempt from personal income tax; the tax base starts at	Q_1
To decide the type of houses a building contractor should construct based on family size	Mode
To compare the incomes of teachers in two school districts	Mean or median
To judge the performance of a worker based on the daily production output	Mean
To determine if the elevator is overloaded based on the peoples' weight	Mean

It is interesting to note that in some problems mean and median both reasonably represent the data. The choice of the measure depends on the type of statistical analysis to be performed on the data.

RANGE, VARIANCE, AND STANDARD DEVIATION

Different data sets may have the same measures of central tendency, but may differ relative to the variability or dispersion of the data. For example

$$10, 10, 10, 10, 10, 10$$

and

$$1, \quad 1, 10, 10, 19, 19$$

both have the same mean and the same median. The former data have no variability whatsoever, whereas the latter is variable. Thus one cannot describe the data only by a measure of central tendency. One also needs a measure of variability to express data. Commonly used measures of variability are: (1) range, (2) variance, and (3) standard deviation (SD).

Range is the difference between the highest and the lowest observations. It is always taken to be positive. It is a crude measure of variability because it does not require all observations in its calculation. Data set with no variability except for one extreme value will have a high range, which defeats the use of the range as a variability measure.

Variance and SD are interrelated concepts and they measure the variability of the data about the mean. Variance is the square of SD, whereas SD is the positive square root of variance. The measures variance and SD make use of all the observations and increase in magnitude with increasing variability in data about the mean.

A data set with smaller variability indicates consistency of the observations. Thus, one data set is more consistent than another data set if it has a smaller variance than the other. Calculation of variance and SD for ungrouped and grouped data sets will be given in the chapter Appendix.

Example 3.13. (Home Run Leaders in National and American Leagues). The home runs hit by home run leaders of National League from 1923 to 1984 have

$$\text{Mean} = 39.7 \qquad \text{SD} = 8.0$$

and the home runs hit by home run leaders of American League from 1923 to 1984 have

$$\text{Mean} = 41.4 \qquad \text{SD} = 8.5$$

The American League home run leaders on an average hit slightly more home runs than the National League leaders. However, the American League players are slightly less consistent than the National League players.

Example 3.14. (Tax Preparers). Let the times (in minutes) taken to complete tax returns by tax preparer A have

$$\text{Mean} = 40 \qquad \text{SD} = 30$$

and B have

$$\text{Mean} = 60 \qquad \text{SD} = 5$$

Here B takes, on an average, a longer time, but is more consistent. Although A takes an average 40 minutes, he is very inconsistent. On some returns he will be very fast, whereas on some returns he will be pretty slow. If one wants to beat the deadline of filing taxes and has only 30 minutes at his disposal, one should think of A, not B, because he is the only hope.

USE OF MEAN AND SD FOR INTERPRETING DATA

For a unimodal distribution, the graph of a curve approximating the histogram is one of the three types indicated in Figures 3.8A, 3.8B, and 3.8C.

In Figure 3.8A, the mean, median, and mode are all the same. The curve is symmetrical about this common measure in the sense that the curve traced to the right of the mean is same as the curve traced to the left of the mean.

In Figure 3.8B, the curve is skewed right. There are many small observations and fewer large observations. The large observations quickly increase the mean compared to the median. In this case, mode $\leq$ median $\leq$ mean.

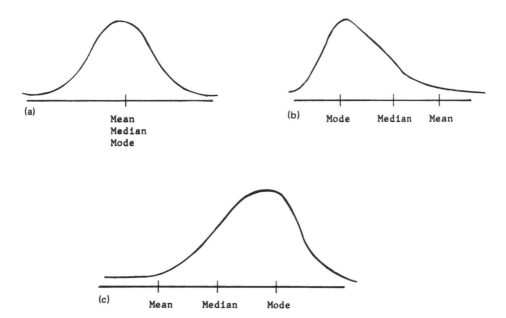

Figure 3.8   Skewed curves (a) symmetrical, (b) right, (c) left.

In Figure 3.8C, the curve is left skewed. There are many large observations and fewer small observations. The smaller observations quickly decrease the mean compared with the median. Here, mean ≤ median ≤ mode.

In all three cases, the median is between the mean and the mode, justifying the name given to it.

A symmetrical curve approximating the histogram with the right height and shape is said to be a normal curve, and the variable associated with that data is said to form a normal distribution. The normal curve is so named because the histograms of the usually studied biological variables have that form. This curve is known in literature from nearly 1720 A.D. Though most of the commonly used continuous variables have a normal distribution, it should not be construed that every continuous variable has a normal distribution. If the phenomenon under discussion yields many observations in the middle portion and tapers off symmetrically on both ends, the variable corresponding to that phenomenon can be reasonably considered to approximate a normal distribution. Statistical tools are available to test whether a data set forms a normal distribution.

Usually data on the following variables are approximately normally distributed:

1. Test scores
2. Heights of adult males of a nation
3. Weights of adult females of a nation
4. Stopping distances of brand X cars at 30 mph
5. Times taken to complete a job
6. Blood pressures of healthy adult males of 30 years age
7. Time between oil changes of cars
8. Commuting distances of students at a large urban university
9. The amount of coffee dispensed at a vending machine
10. Gestation period in women

The data on the following variables usually do not follow the normal distribution curve.

1. Personal income of people in a community
2. Life length of electric bulbs
3. Number of brothers (not including step-brothers of men in a community)
4. The number of defectives based on small samples in a production line
5. The number of bedrooms in houses in a community

When data follows an exact or approximate normal distribution, the percentage of observations in different intervals are schematically shown in Figure 3.9.

For a large number of observations based on a normal distribution, the range is nearly 6(SD). Thus, without calculating the SD, one can guess the SD of the data as approximately range/6. With a small number of observations, the smallest probable observation becomes the mean − 2(SD) and the largest probable observation becomes the mean + 2(SD) and here one guesses the SD as range/4. These are only guess values and one needs to calculate exact values for the SD by using the formulae given in the Appendix.

Some examples illustrating these ideas about the normal distribu-

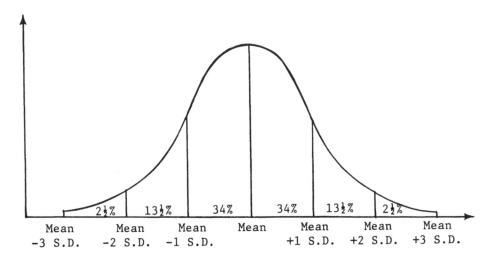

Figure 3.9 Approximate percentage of observations under a normal curve.

tion will be discussed in the following concluding examples of this chapter.

Example 3.15. (Test Scores). In a national examination, let 20% be the lowest score obtained by any student and let 90% be the highest score obtained by any student. Because test scores are usually normally distributed and for a normal distribution the mean is in the middle of a curve, it follows that the

$$\text{Mean of the test scores} = \frac{20 + 90}{2} = 55$$

Because the range for the data is $90 - 20 = 70$, the SD is approximately given by $70/6 = 11.7$. One then forms Figure 3.10.

From Figure 3.10 it is understood that nearly 2.5% of the students got scores between 19.9 and 31.6, nearly 13.5% between 31.6 and 43.3; nearly 34% between 43.3 and 55; nearly 34% between 55 and 66.7; nearly 13.5% between 66.7 and 78.4; and nearly 2.5% between 78.4 and 90.1.

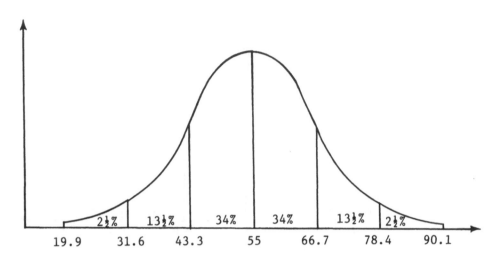

Figure 3.10 Approximate percentages for test score data.

Example 3.16. (National Basketball Association Scoring Leaders). The averages of NBA scoring leaders from *The World Almanac and Book of Facts, 1985,* p. 811 are

23.2	21.0	28.3	27.4	28.4	25.4	22.3
24.4	22.7	25.7	25.6	27.8	29.2	37.9
38.4	50.4	44.8	36.5	34.7	33.5	35.6
27.1	28.4	31.2	31.7	34.8	34.0	30.6
34.5	31.1	31.1	27.2	29.6	33.1	30.7
32.3	28.4	30.6				

For this data, the mean = 30.8 and SD = 5.9. The actual percentage of observations and theoretical percentage of observations assuming normality are indicated in Table 3.14. One can observe the reasonable proximity of the actual and theoretical percentages. The theoretical percentages are given on the premise that the average scores are normally distributed. The agreement between the actual and theoretical percentages of Table 3.14 can be tested by methods found in Chapter 7 using actual and expected frequencies in different classes (see Review Exercise 60).

Table 3.14 Actual and Theoretical
Percentages

Class	Actual percentage	Theoretical percentage
13.1–19.0	0	2.5
19.0–24.9	13.16	13.5
24.9–30.8	42.11	34.0
30.8–36.7	34.21	34.0
36.7–42.6	5.26	13.5
42.6–48.5	2.63	2.5
≥48.5	2.63	0

REFERENCES

Bishop, Y.M.M., Fienberg, S.E., and Holland, P.W. (1975). *Discrete Multivariate Analysis: Theory and Practice*. MIT Press, Cambridge.

Bishop, Y.M.M. (1969). Full contingency tables, logits, and split contingency tables. *Biometrics* 25. 119–128.

Ryan, B.F., Joiner, B.L., and Ryan, Jr. T.A. (1985). *Minitab Handbook*. Duxbury Press.

Snedecor, G.W. (1947). *Proc. Int. Statist. Conf.* 3. 440.

Snedecor, G.W. and Cochran, W.G. (1980). *Statistical Methods*. The Iowa State University Press.

Tukey, J.W. (1977). *Exploratory Data Analysis*. Addison Wesley, Reading, Mass.

APPENDIX

Forming Frequency Tables and Histograms

The procedure of forming frequency tables and histograms is discussed in the following examples:

Example 3A.1.   The raw data of 40 observations are

$$8, 36, 24, 28, 34, 12, 17, 21, 9, 27,$$
$$31, 12, 17, 13, 24, 26, 30, 15, 7, 19,$$
$$6, 18, 21, 24, 34, 9, 15, 18, 29, 26,$$
$$13, 15, 18, 18, 24, 29, 24, 17, 16, 22,$$

and one is interested in forming a frequency table and a histogram for the data. The following stepwise procedure can be followed:

1. Determine the range, R, of the data, which is the difference of the largest and smallest observations. Here $R = 36 - 6 = 30$.
2. Determine the number of classes, k, for the frequency table. This number is arbitrary and is the choice of the investigator. Usually 5 to 15 classes are recommended. Depending upon the nature of the data, R, and the number of observations, the number of desirable classes in a frequency table can be determined. Here $k = 6$.
3. Determine the class length, C, as a number slightly higher than R/k, to the same degree of accuracy that the observations are given. Here $R/k = 30/6 = 5$, and because the data is integral, the class length C will be taken as 6.
4. Determine the lowest-class boundary of the beginning class as the left-side starting number rounded off to the smallest observation in the data. Here the smallest observation is 6 and numbers 5.5 and above and below 6.5 are rounded to 6. Thus the lowest class boundary of the beginning class will be taken as 5.5.
5. Form classes and count the frequency in each class. The upper-class boundary of the first class is obtained by adding C to the decided lower-class boundary at item 4. The upper class boundary of the first class is taken as the lower class boundary of the second class. The upper class boundary of the second class is obtained by adding C to its lower-class boundary. The upper-class boundary of the second class is then taken as the lower-class boundary of the third class and the process continued until k classes are formed. The data is then counted into the classes by making tally marks in groups of five. Table 3A.1 is so constructed for the present data.
6. A histogram is then plotted by taking the variable on the X axis and the frequency on the Y axis. With equal class lengths, bars of heights proportional to frequencies are constructed over each class. In the case of unequal class lengths in a histogram, bars are

Table 3A.1   Frequency Table

Class	Frequency		Percentage
5.5–11.5	৷৷৷৷	= 5	$\frac{5}{40}(100) = 12.5$
11.5–17.5	৷৷৷৷ ৷৷৷৷ |	= 11	$\frac{11}{40}(100) = 27.5$
17.5–23.5	৷৷৷৷ ৷৷৷	= 8	$\frac{8}{40}(100) = 20.0$
23.5–29.5	৷৷৷৷ ৷৷৷৷ |	= 11	$\frac{11}{40}(100) = 27.5$
29.5–35.5	||||	= 4	$\frac{4}{40}(100) = 10.0$
35.5–41.5	|	= 1	$\frac{1}{40}(100) = 2.5$
Total		40	100.0

constructed over each class interval the heights of which are the frequencies per each unit length in that class. The areas of the bars in a histogram are proportional to the frequencies in the classes. The histogram for the data at hand is illustrated in Figure 3A.1.

Example 3A.2.   Consider the following data set of 25 observations.

$$-1.6, 2.1, -1.4, \quad 1.5, \quad 0.4, -0.7, -1.2, 1.7,$$
$$1.3, 2.2, -2.1, -1.5, \quad 1.3, \quad 0.6, -0.3, 0,$$
$$1.8, 1.3, \quad 0.4, \quad 2.3, -1.5, -2.1, -1.1, 1.8, 0.2$$

The stepwise procedure discussed with the previous example yield the following:

1. $R = 2.3 - (-2.1) = 4.4$
2. $k = 5$
3. $R/k = 4.4/5 = 0.88$. Take $C = 0.9$ because data is given to tenths.

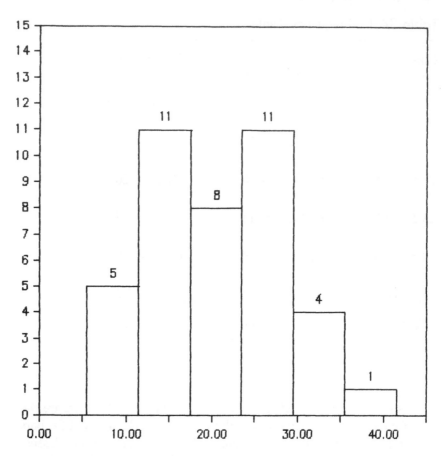

Figure 3A.1 Histogram for the data of Table 3A.1.

4. Because the lowest observation is −2.1 and the numbers −2.15 and above, and below −2.05 are rounded to −2.1, the lowest class boundary of the initial class is −2.15.
5. The frequency table can then be formed as in Table 3A.2.
6. The histogram can then be drawn as in Figure 3A.2.

Example 3A.3.   Consider the drainage area data of Table 3.3. The classes are of unequal length. Hence, to construct the histogram, one expresses the frequency on a comparable basis for all classes. The frequency for every 10,000 in each class can be easily calculated as 3,

Table 3A.2 Frequency Table

Classes	Frequency	Percentage
$(-2.15)-(-1.25)$	卌 丨 = 6	$\frac{6}{25}(100) = 24.0$
$(-1.25)-(-0.35)$	‖‖ = 3	$\frac{3}{25}(100) = 12.0$
$(-0.35)-(+0.55)$	卌 = 5	$\frac{5}{25}(100) = 20.0$
$0.55 - \quad 1.45$	‖‖‖ = 4	$\frac{4}{25}(100) = 16.0$
$1.45 - \quad 2.35$	卌 ‖ = 7	$\frac{7}{25}(100) = 28.0$
Total	25	100.0

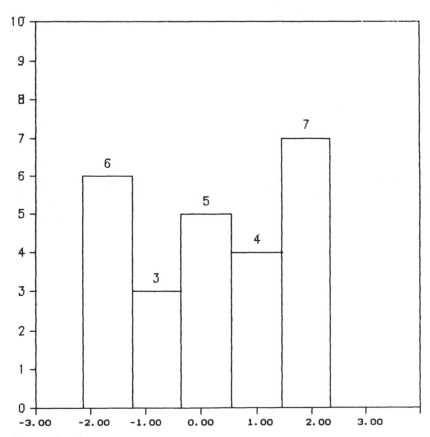

Figure 3A.2 Histogram for the data of Table 3A.2.

10, 2, 5, 0.4, 0.2, 0.12, and 0.02 respectively. The histogram can then be drawn by constructing bars of these heights for the classes indicated in Table 3.3.

Mean, Variance, SD, and Median for Ungrouped Data

The population mean is denoted by μ and the sample mean is denoted by $\bar{X}$. The mean is a simple arithmetic average of all observations. Let $X_1, X_2, \ldots, X_n$ be a sample of n observations, then the mean is given by

$$\bar{X} = \frac{X_1 + X_2 + \cdots + X_n}{n} \tag{3A.1}$$

The population variance is denoted by σ^2 and sample variance by s^2. For a sample,

$$s^2 = \frac{1}{n-1} \{(X_1 - \bar{X})^2 + (X_2 - \bar{X})^2 + \cdots + (X_n - \bar{X})^2\}$$

$$= \frac{(X_1^2 + X_2^2 + \cdots + X_n^2) - \dfrac{(X_1 + X_2 + \cdots + X_n)^2}{n}}{n-1} \tag{3A.2}$$

The second expression for s^2 given above follows from the first expression through some algebraic manipulation. The first expression for s^2 indicates that it measures the variability of the data about $\bar{X}$. The sample SD denoted by s is the positive square root of s^2 given by (3A.2). The population SD is denoted by σ.

   To obtain the median, the data is first ranked, that is, arranged in increasing order. The median is the $\dfrac{n+1}{2}$th observation in the ranked data if n is odd and it is the average of the $\dfrac{n}{2}$th and $\left(\dfrac{n}{2} + 1\right)$th observations when n is even.

   Example 3A.4.   Consider the data

$$6, 7, 8, 10, 4$$

Clearly, $n = 5, 6 + 7 + 8 + 10 + 4 = 35$, and $6^2 + 7^2 + 8^2 + 10^2 + 4^2 = 265$. Hence

$$\bar{X} = \frac{35}{5} = 7$$

$$s^2 = \frac{265 - (35^2/5)}{5 - 1} = 5$$

$$s = \sqrt{5} = 2.24$$

The data arranged in increasing order is 4, 6, 7, 8, 10. Here n is odd and hence the $\left(\frac{5 + 1}{2} = 3\right)$rd observation, 7, is the median of the data.

Example 3A.5. Consider the data

$$15, 26, 13, 29, 8, 6$$

Clearly $n = 6$, $15 + 26 + 13 + 29 + 8 + 6 = 97$, and $15^2 + 26^2 + 13^2 + 29^2 + 8^2 + 6^2 = 2011$. Hence

$$\bar{X} = \frac{97}{6} = 16.17$$

$$s^2 = \frac{2011 - (97^2/6)}{6 - 1} = 88.57$$

$$s = 9.41$$

The data arranged in increasing order is 6, 8, 13, 15, 26, 29. Here n is even, and hence, the average of the $\left(\frac{6}{2} = 3\right)$rd and $\left(\frac{6}{2} + 1 = 4\right)$th observations, $\frac{13 + 15}{2} = 14$, is the median of data.

Example 3A.6. Consider the data

$$-0.11, 0.21, -0.39, 0.42, -0.23, 0.16, 0$$

Clearly $n = 7$, $(-0.11) + (0.21) + (-0.39) + (0.42) + (-0.23) +$

$(0.16) + 0 = .06$, $(-0.11)^2 + (0.21)^2 + (-0.39)^2 + (-0.23)^2 + (0.16)^2 = 0.4632$. Hence

$$\bar{X} = \frac{0.06}{7} = 0.01$$

$$s^2 = \frac{0.4632 - (0.06^2/7)}{7 - 1} = 0.08$$

$$s = 0.28$$

The data in increasing order is $-0.39, -0.23, -0.11, 0, 0.16, 0.21,$ 0.42. Here n is odd and hence the $\left(\frac{7 + 1}{2} = 4\right)$th observation, 0, is the median.

If an observation larger than $\bar{X}$ is removed from a data set, the mean of the resulting data set will be smaller than $\bar{X}$. If an observation smaller than $\bar{X}$ is removed from a data set, the mean of the resulting data set will be more than $\bar{X}$. If a new observation larger than $\bar{X}$ is added to a data set, the mean of the resulting data will be more than $\bar{X}$. Similarly, if a new observation smaller than $\bar{X}$ is added to a data set, the mean of the resulting data set will be less than $\bar{X}$.

Weighted Mean

When there are k different data sets with $n_1, n_2, \ldots, n_k$ observations with respective means $\bar{X}_1, \bar{X}_2, \ldots, \bar{X}_k$, one can combine all these k sets into one large set and determine the weighted mean of the combined group as $\bar{X}_w$, the subscript w denoting weighted mean, given by the formula:

$$\bar{X}_w = \frac{n_1\bar{X}_1 + n_2\bar{X}_2 + \cdots + n_k\bar{X}_k}{n_1 + n_2 + \cdots + n_k} \qquad (3A.3)$$

Clearly $\bar{X}_w$ is larger than the smallest individual mean and is smaller than the largest individual mean.

Example 3A.7.   Let there be three sections of a first course in statistics with enrollment of 25, 30, and 40 students. Let the average test scores for the three sections in the final test are 80, 75, and 60. The mean for the combined group of $25 + 30 + 40 = 95$ students is given by (3A.3) and is

$$\bar{X}_w = \frac{(25)(80) + (30)(75) + (40)(60)}{25 + 30 + 40} = 70$$

Mean for Grouped Data

Consider a frequency table with k classes. Let $f_1, f_2, \ldots, f_k$ be the frequencies in the k classes and let $x_1, x_2, \ldots, x_k$ be the middle points for the k classes. Because the f_i observations are assumed to have been equally spaced over the ith class and the mean of equally spaced observations occurs at the center, the mean of f_i observations in ith class is x_i. In view of the foregoing discussion, the mean of all observations is given by the formula

$$\bar{x} = \frac{(f_1)(x_1) + (f_2)(x_2) + \cdots + (f_k)(x_k)}{n} \qquad (3A.4)$$

and the variance is given by the formula

$$s^2 = \{[(f_1)(x_1^2) + (f_2)(x_2^2) + \cdots + (f_k)(x_k^2)] \\ - [(f_1)(x_1) + (f_2)(x_2) + \cdots + (f_k)(x_k)]^2/n\}/(n - 1) \quad (3A.5)$$

Example 3A.8. Consider the frequency Table 3A.3. The third col-

Table 3A.3

Class	f_i	x_i
50–60	4	$(50 + 60)/2 = 55$
60–70	6	$(60 + 70)/2 = 65$
70–80	5	$(70 + 80)/2 = 75$
80 90	3	$(80 + 90)/2 = 85$
90–100	2	$(90 + 100)/2 = 95$
Total	20	

umn of the table is obtained by averaging the lower and upper class boundaries in each class. Then from (3A.4)

$$\bar{x} = \frac{(4)(55) + 6(65) + 5(75) + 3(85) + 2(95)}{20} = \frac{1430}{20}$$

$$= 71.5$$

$$s^2 = \frac{[(4)(55^2) + (6)(65^2) + (5)(75^2) + (3)(85^2) + (2)(95^2)] - \dfrac{1430^2}{20}}{19}$$

$$= 160.7895$$

Example 3A.9.   Consider the frequency Table 3A.4. From (3A.4),

$$\bar{X} = \frac{(2)(-0.15) + (3)(-0.05) + (5)(0.1) + (10)(0.3) + (20)(0.7)}{40}$$

$$= \frac{17.2}{40}$$

$$= 0.43$$

$$s^2 = \frac{[(2)(-0.15)^2 + 3(-0.05)^2 + 5(0.1)^2 + 10(0.3)^2 + 20(0.7)^2] - \dfrac{(17.2)^2}{40}}{39}$$

$$= 0.0873$$

Table 3A.4

Class	f_i	x_i
(−0.2)–(−0.1)	2	$((-0.2) + (-0.1))/2 = -0.15$
(−0.1)– 0	3	$((-0.1) + 0)/2 = -0.05$
0 − 0.2	5	$(0 + 0.2)/2 = 0.1$
0.2 − 0.4	10	$(0.2 + 0.4)/2 = 0.3$
0.4 − 1.0	20	$(0.4 + 1.0)/2 = 0.7$
Total	40	

Minitab Computer Package

Minitab is a computer package in which the user speaks to the computer in simple English commands. While it is possible to do even complicated problems with this language, simple commands providing direct results will be discussed in this book.

Depending on the type of the computer accessible to the reader, with appropriate commands the user should LOGIN to get to MINITAB. In Minitab, the computer prompts MTB > asking the user to give a command. The data is entered in columns labeled C1, C2, . . . , depending on the number of columns in which the data is available, through the READ command. Now Minitab expects data, and prompts with DATA >. The data is then entered line after line, returning the carriage at the end of each line, and the command END finishes the input of the data. If any mistake is done in entering the data, it can be corrected through the LET command. A wrong entry in the fourth row of column two can be corrected by typing the correct value after equal sign in

MTB > LET C2(4) =

The necessary commands to run the statistical programs as needed will be presented in this book. For clarity, the words typed by the user will be shown in lower case letters and the characters that the computer prints out will be in capital letters.

Let the reader who is interested in using the Minitab package get descriptive statistics, histogram, stem-and-leaf plot, and box plot (or box-and-whisker plot) for the data given in Examples 3A.1 and 3A.2. For the data of Example 3A.1 the following are the necessary commands typed by the reader and the output provided by the computer.

```
MTB>read c1
DATA>8
DATA>36
DATA>24
DATA>28
DATA>34
DATA>12
DATA>17
DATA>21
```

```
DATA>9
DATA>27
DATA>31
DATA>12
DATA>17
DATA>13
DATA>24
DATA>26
DATA>30
DATA>15
DATA>7
DATA>19
DATA>6
DATA>18
DATA>21
DATA>24
DATA>34
DATA>9
DATA>15
DATA>18
DATA>29
DATA>26
DATA>13
DATA>15
DATA>18
DATA>18
DATA>24
DATA>29
DATA>24
DATA>17
DATA>16
DATA>22
DATA>end
    40 ROWS READ
```

```
MTB>describe cl
               Cl
N              40
MEAN        20.15
MEDIAN      18.50
TMEAN       20.08
STDEV        7.87
SEMEAN       1.24
MAX         36.00
MIN          6.00
Q3          26.00
Q1          15.00
MTB>histogram cl

       Cl

     MIDDLE OF      NUMBER OF
     INTERVAL       OBSERVATIONS
            4       0
            8       5     *****
           12       4     ****
           16       7     *******
           20       7     *******
           24       6     ******
           28       6     ******
           32       2     **
           36       3     ***

MTB>stem-and-leaf cl

     STEM-AND-LEAF DISPLAY OF Cl
     LEAF DIGIT UNIT =      1.0000
     1 2 REPRESENTS 12.

         2      +0S     67
         5      +0.     899
         5      1*
         9      1T      2233
        12      1F      555
        16      1S      6777
        (5)     1.      88889
```

```
19          2*      11
17          2T      2
16          2F      44444
11          2S      667
 8          2.      899
 5          3*      01
 3          3T
 3          3F      44
 1          3S      6
```

```
CONTINUE?y
MTB>boxplot c1
```

```
ONE HORIZONTAL SPACE = .70E+00
FIRST TICK AT        7.000
```

In the descriptive statistics obtained on the printout, SEMEAN is $s/\sqrt{n}$ and TMEAN is the mean of 90% of the observations obtained after deleting the lower 5% and upper 5% observations. Max and Min respectively denote the largest and smallest observations in the data.

The second column in the stem-and-leaf display gives the stems, and the entries in the third column are the leaves. Multiple lines are entered for some stems. The entries in the first column is a frequency count of the leaves on the stem. The line in which the frequency is shown in parenthesis contains the median.

Using subcommands, one may adjust the histogram with respect to the class length and beginning value.

For the data of Example 3A.2 the following are the necessary commands and the output provided by the computer.

```
MTB>read c1
DATA>1.6
DATA>2.1
```

```
DATA>-1.4
DATA>1.5
DATA>.4
DATA>-.7
DATA>-1.2
DATA>1.7
DATA>1.3
DATA>2.2
DATA>-2.1
DATA>-1.5
DATA>1.3
DATA>.6
DATA>-.3
DATA>0
DATA>1.8
DATA>1.3
DATA>.4
DATA>2.3
DATA>-1.5
DATA>-2.1
DATA>-1.1
DATA>1.8
DATA>.2
DATA>end
     25 ROWS READ

MTB>describe c1

                 C1
N                25
MEAN           0.22
MEDIAN         0.40
TMEAN          0.23
STDEV          1.47
SEMEAN         0.29
MAX            2.30
MIN            2.10
Q3             1.60
Q1            -1.30
MTB>histogram c1
```

```
C1

MIDDLE OF      NUMBER OF
INTERVAL       OBSERVATIONS
   -2.0           2     **
   -1.5           4     ****
   -1.0           2     **
   -0.5           2     **
    0.0           2     **
    0.5           3     ***
    1.0           0
    1.5           5     *****
    2.0           4     ****
    2.5           1     *
```

MTB>stem—and—leaf c1

```
STEM—AND—LEAF DISPLAY OF C1
      LEAF DIGIT UNIT = .1000
1 2 REPRESENTS 1.2

   2        -2*     11
   5        -1.     655
   8        -1*     421
   9        -0.     7
  10        -0*     3
  (4)       +0*     0244
  11        +0.     6
  10         1*     333
   7         1.     5788
   3         2*     123
```

MTB>boxplot c1

```
                    --------------------------
      --------------I              +       I------------
                    --------------------------
      -+-------------+------------+-------+---------+---
```

ONE HORIZONTAL SPACE = .10E+00
FIRST TICK AT   -2.000

When only one column of data is entered, it is more efficient to enter the data by a SET command and type the data in a single line or several lines using a comma or space between entries. To be consistent with the data entry in other chapters, in the above two examples data were entered through a READ command.

After completing the Minitab, the user should exit from Minitab with a STOP command and LOGOUT from the main computer.

EXERCISES

1. Justify the appropriateness of the indicated measures of central tendencies for the settings given on p. 66.
2. Justify that the data on the variables given on p. 70 produce a normal histogram.
3. Find the median, mean, variance and SD for the following data sets
   (a) 100, 80, 70, 69, 75, 94, 82
   (b) −0.1, 0.3, −0.7, −0.6, 0.3, 0.7
   (c) 8.100, 9.750, 1.341, 2.128
   (d) 45, 26, 38, 61, 29, 53, 81
   (e) 4, 8, 6, 5, 1, 3
4. Find the mean drainage area using the data of Table 3.3.
5. Find the mean number of operating school districts using the data of Table 3.4.
6. Form a suitable frequency table and histogram for the following data:

   12, 20, 18, 25, 20, 5, 15, 12, 12, 20, 6, 15,
   8, 12, 35, 40, 15, 7, 10, 8, 20, 4, 25, 20, 7

7. Form a suitable frequency table and histogram for the following data:

   365, 187, 219, 225, 240, 122, 278, 406, 63, 231,
   31, 284, 201, 61, 645, 250, 52, 425, 151, 257

8. If test scores are normally distributed with a mean of 70 and a SD of 10, what percentage of students do you expect to score between 60 and 80?

9.   Stopping distances for cars at 30mph is 50 ft with a SD of 5 ft. Is it probable to stop a car at 30mph within 30 ft after applying the brakes?

10. The chest measurements of adult males are normally distributed with a mean of 41 inches and a SD of 2 inches. What is the range of chest measurements?

Answers

(3a) 80, 81.43, 138.62, 11.77; (3b) 0.1, −0.02, 0.31, 0.55; (3c) 1.73, 3.86, 13.66, 3.70; (3d) 45, 47.57, 372.62, 19.30; (3e) 4.5, 4.5, 5.9, 2.43; (4) 146, 333.3; (5) 311.76; (8) 68%; (9) no; (10) 12 inches, from 35 to 47 inches.

4

Simple Inferences that Can Be Drawn for One Population

The main purpose of statistics is to enable a person to make inferences about the unknown population from the known sample data. This is the process of inductive reasoning comparable to a detective reconstructing a crime scene based on the available clues. Although in novels the detective never goes wrong, in real life there is some chance that the detective can be wrong and the innocent person will be charged with the crime. In fact, the conclusions based on inductive logic are never 100% valid and there is some probability that they are incorrect. To understand this chance mechanism, a student of statistics must have some knowledge of probability theory and it will be introduced in the next section.

The values characterizing the population observations are called *parameters*. These are, for example, the population mean denoted by μ, the population variance denoted by σ^2, and the population proportion p characterizing some given property. Analogous values characterizing the sample observations are called *statistics*. These are, the sample mean denoted by $\bar{X}$, the sample variance denoted by s^2, and the sample proportion $\hat{p}$, which will be read as "p hat." Estimation and tests of hypotheses concerning the parameters will be based on sample statistics and will be discussed in the subsequent sections.

PROBABILITY

Statements such as "I will probably get an A grade in statistics," "Most probably it will rain tomorrow," or "It is more probable to win at a black-jack table than at a slot machine in a casino," are often heard. The ideas involved here are vague and do not specify the exact chances. In statistics and in all types of inferential problems one is interested in ascertaining the exact chance of occurrence of any specified event.

Probability, on a scale of 0 to 1, indicates the chance of an event under consideration happening. The simplest example of probability is the weatherman's report on a television station. When there is a 40% chance of rain for tomorrow, it implies that on days with a weather profile (temperature, humidity, barometric pressure, cloud formation, etc.) similar to tomorrows, it rained 40% of the time. It does not tell you with surety that it will rain tomorrow or that it will not rain tomorrow. Different reporters interpret it as a rainy day or an occasional shower day, or a cloudy day depending on their personal outlook on nature. When there is a 1 in a million chance of winning a grand prize in a lottery, it means that there is one winning ticket in a million tickets sold. It does not provide a guarantee that the buyer of a lottery ticket is a winner or a loser.

Probability theory was originally developed to answer gambling problems and slowly found applications into many subject fields. There are three main methods of calculating probability: (1) subjective method, (2) relative frequency method, and (3) mathematical method.

The subjective method of determining probability depends on the individual assessing the chances. Two individuals need not necessarily determine the same probability for a given event. The probability that there will be a nuclear war at the end of this century can be determined only by the subjective method. It is conceivable to think that no two individuals may give the same subjective probability that there will be a nuclear war at the end of this century. One may think it to be 0.9, whereas another may assign it 0.5.

The relative frequencies that were introduced in the last chapter are, in a sense, probabilities if they are based on a large number of observations. In fact, probability of an event is a long run stability of a relative frequency. If a baseball player gets 150 hits in 500 times at bat, his relative frequency of making a hit is 150/500 = 0.3. Next time when that player is batting, he has a 0.3 probability of making a hit. If a teacher gives 15% A grades in a class, the probability that a given student gets an A grade is 0.15. If an operation is successful for 700 patients out of 1000 operated patients, the relative frequency of success for the operation is 700/1000 = 0.7. The probability for success of the operation for the person on the operating table is 0.7. It may be noted that the relative frequencies in the above examples should be based on a large number of observations in order that they be considered as probabilities.

The mathematical formulation of probability is an axiomatic approach hinged on the idea of a sample space, and the probabilities are weights attached to the points of the sample space (see Appendix). The following are some illustrations of probabilities calculated by the ideas to be developed in the Appendix:

Example 4.1. (Throwing a Pair of Fair Dice). The probability of throwing a double with a pair of fair dice is 6/36 = 1/6 (=16.7%). This implies that, on an average, one gets a double for every six throws. The probability of throwing a total of 7 or 11 is 2/9 = (=22.2%). Thus, on an average, one gets either a 7 or an 11, twice every nine throws.

Example 4.2. (Three-Children Families). The probability that all of the three children are of the same sex in three-children families is 1/4 (=25%). Thus, on an average, one of every four three-children families has either all boys or all girls. The probability that there is at least one boy in a three-children family is 7/8 (=87.5%). This means that, on an average, seven of every eight three-children families have at least one boy (i.e., either 1, 2, or 3 boys).

Example 4.3. (Five-Card Poker Game). The probability of getting a royal flush (ace, king, queen, jack, and ten of the same suit) is 1/649,740. On an average, one of every 649,740 poker hands is a royal flush. The probability of getting a straight flush (five cards in a sequence, all of the same suit excluding a royal flush) is 9/649,740. On an average, nine of every 649,740 poker hands is a straight flush.

A straight flush is more probable than a royal flush. The probability of getting a four-of-a-kind (four cards of the same face value) is 1/4165. On an average, one of every 4165 poker hands is a four-of-a-kind. The probability of getting a full house (three cards of the same face value and the other two cards of another common face value) is 6/4165. On an average, six of every 4165 poker hands is a full house. The probability of a flush (five cards of the same suit, excluding cards in a sequence) is 1277/649,740. On an average, 1277 of every 649,740 poker hands is a flush.

Example 4.4. (Bridge Hands). At a bridge table, the probability that each player holds an ace is 2197/20,825 (=10.5%). On an average, in 2197 deals of 20,825 deals, each of the four players will have an ace. The probability that a player has all four aces is 44/4165 (=1.1%). On an average, in 44 of every 4165 deals, one of the four players will have all four aces. Thus it is more likely to see each player getting an ace rather than one player having all aces.

Events with small probability are usually unrealized in a single trial. If the weather man predicts a 10% chance of rain tomorrow, people may pay no attention to "rain tomorrow" and may even plan outdoor parties. However, when there is an 80% chance of rain for tomorrow, people take all precautions against rain for tomorrow. Note that even though it is unlikely to rain on a given day when there is a 10% chance of rain, it will rain, on an average, one out of every ten of those days predicted as having a 10% chance of rain. It is highly improbable to see a royal flush in one poker hand; but, one may get a royal flush if poker is continuously played.

One needs to distinguish between *impossible* and *improbable* events. An impossible event has zero probability, while an improbable event has a small probability. An improbable event does not mean that it is impossible to happen. A student who thinks it improbable to receive an A grade in a statistics course may very well receive an A grade in the course. Improbable events are those that are very unlikely to occur, whereas impossible events are the ones which cannot occur under any circumstances.

The events for which probabilities are evaluated are not necessarily quantitative. By assigning suitable numerical values to the events, one defines random variables. The random variables are usually designated by capital Roman letters such as X, Y, Z, U, V, W. A random variable

X is either discrete or continuous. It is discrete if it takes distinct values, that can be counted; and it is continuous if it takes values in an interval with no gaps (also see Chap. 1).

A discrete random variable will be completely described by a *probability distribution function* (pdf). For a random variable X taking values $x_1, x_2, \ldots, x_k$ with corresponding probabilities $p_1, p_2, \ldots, p_k$, the pdf will be shown as

X	x_1	x_2	$\cdots$	x_k
$p_X(x)$	p_1	p_2	$\cdots$	p_k

where the top line of the table provides the values taken by the random variable and the bottom line gives the corresponding probabilities. The sum of the bottom line probabilities must be 1. Given a pdf of a random variable X, one calculates the average and variance of the random variable X. However, as a random variable describes a general phenomenon, the average value is called the expected value of the random variable and is denoted by $E(X)$ or μ_X. The variance will be denoted by σ_X^2. The subscript X for μ and σ^2 can be used or suppressed as needed. The calculation of μ and σ^2 will be discussed in the Appendix. Some examples of pdfs will now be given.

Example 4.5. (More on Throwing a Pair of Fair Dice). Let X be the sum of the two face values when a pair of fair dice are thrown. Then X has the following pdf:

X	2	3	4	5	6	7	8	9	10	11	12
$p_X(x)$	$\dfrac{1}{36}$	$\dfrac{2}{36}$	$\dfrac{3}{36}$	$\dfrac{4}{36}$	$\dfrac{5}{36}$	$\dfrac{6}{36}$	$\dfrac{5}{36}$	$\dfrac{4}{36}$	$\dfrac{3}{36}$	$\dfrac{2}{36}$	$\dfrac{1}{36}$

From this table one observes that a total of 2 can be obtained, on an average, once for every 36 throws, a total of 3 can be obtained, on an average, twice for every 36 throws, a total of four can be obtained on an average thrice out of every 36 throws, and so on. Here $\mu_X = 7$. Although it is possible to obtain any of 2, 3, $\ldots$, or 12 as the sum of the two face values when a pair of dice are thrown, on an average, a sum of 7 is expected when a pair of dice are thrown.

Example 4.6. (Insurance Payoff). Suppose an insurance company receives annual claims on 10,000 dollar group life policies 1% of the time. Defining X as the company's payoff, the pdf of X is

X	0	$10,000
$p_X(x)$	0.99	0.01

Here $\mu_X = \$100$. Thus, although on a policy the company either pays nothing or 10,000 dollars, on an average, it pays 100 dollars. If the company wants to make a profit of 50 dollars on a policy, the annual premiums for a 10,000 dollar group life policy will be fixed at $100 + \$50 = \$150$.

Example 4.7. (Should this Book Be Published?) Before taking a business venture, the management usually conducts a market survey to determine the expected net profit (=net revenue − cost) and if it is reasonable, decides on undertaking the project. For simplicity, let the pdf of the net profit X as determined by a publisher for this book, be the following:

X	$100,000	$50,000	$10,000	0	−$30,000
$p_X(x)$	0.05	0.10	0.40	0.30	0.15

Here $\mu_X = 9500$ and the expected net profit in publishing this book is 9500 dollars. It may be noted that the publisher will not make a net profit of 9500 dollars on this book. He may make a profit of 100,000, 50,000, or 10,000 dollars. There is also the possibility that he may break even or even incur a loss of 30,000 dollars. From undertaking publication of similar works, he makes, on an average, a net profit of 9500 dollars on each title, and thus the publisher should decide whether or not this is an acceptable profit margin for his investment.

There are several standard models for evaluating probabilities of a discrete random variable. Of them, the simplest and most commonly used distributions are binomial, Poisson, and hypergeometric. The settings and the probability calculations under these three distributions will be discussed in the Appendix.

A continuous random variable X will be described by a probability density function $f_X(x)$. There is a zero probability that a continuous random variable assumes any specified value. One draws a graph (or curve) of the probability density function and evaluates the probabilities by finding the appropriate areas between the graph of the density function and the X-axis. One of the widely used distributions (or models) of a continuous random variable X is the normal distribution. The density function of a normal distribution depends on the mean μ and the variance σ^2 of the random variable; here μ is called the location parameter and σ^2 is called the scale parameter. Consider the graphs of two normal distributions with the same σ^2 and different μs. They will have the same shape, but will be centered (or located) at different spots on the X-axis. This situation is illustrated in Figure 4.1. Again, consider the graphs of two normal distributions with same μ and different σ^2s. They will be centered (or located) at the same position, but will have different shapes. This situation is illustrated in Figure 4.2.

To evaluate probabilities using a normal distribution one needs to evaluate the appropriate areas under the graph and this can be easily and conveniently handled by constructing suitable tables or charts. However, there are infinitely many normal distributions corresponding to different μ and σ^2 values. Thus it becomes necessary to unify the

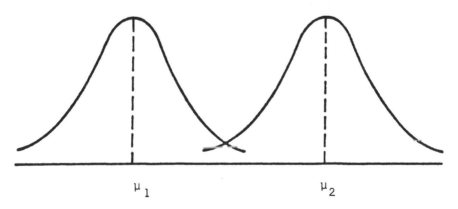

Figure 4.1 Normal curves with $\mu_1 < \mu_2$ and same SD.

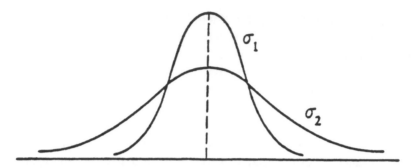

Figure 4.2   Normal curves with same mean and $\sigma_1 < \sigma_2$.

normal distribution by introducing a standard normal random variable
Z given by

$$Z = \frac{X - \mu_X}{\sigma_X}$$

Z is the random variable measuring the original random variable from
the mean in terms of the standard deviation (SD) units. The mean and
variance of the standard normal random variable is given by $\mu_Z = 0$,
$\sigma_Z^2 = 1$. The graph of a standard normal random variable Z is illustrated in Figure 4.3 and the salient features of this graph include

1. The graph is bell-shaped (or mound-shaped) centered at zero en-
    tirely lying above the axis.
2. The graph is symmetric about 0 in the sense that the curve is
    drawn identically to the right and to the left of 0.
3. The total area under the curve is 1.
4. Though theoretically the curve extends from $-\infty$ to $+\infty$, prac-
    tically the curve shapes up from -3 to $+3$.
5. There is a heavier concentration of the area in the middle, tapering
    off at both right and left tails.

Analogous to the definition of the standard normal random variable, every observation can be measured from the mean in terms of the
standard deviation units and this measure is called the z-score (or
standard score) of the measurement. The z-score is a unit-free mea-

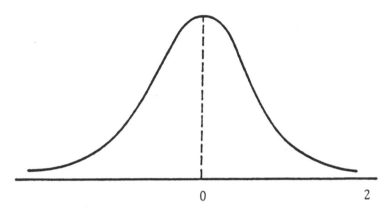

Figure 4.3   The standard normal curve.

surement, and it tells how many standard deviations away the observation is from the mean.

Example 4.8. (Test Scores). An instructor taught two sections of an elementary statistics course. The test scores of Section 1 are normally distributed with mean $\mu_1 = 80$ and the SD $\sigma_1 = 5$. John, a student, in that section received a score of 87. The test scores of Section 2 are also normally distributed with $\mu_2 = 82$ and SD $\sigma_2 = 2$. A student of Section 2, Bob, received a test score of 86. Although John got a higher test score than Bob, it cannot be immediately concluded that John performed any better than Bob. The z-score of John is $(87 - 80)/5 = 1.4$ and the z-score of Bob is $(86 - 82)/2 = 2$. John is 1.4 SDs above the mean; whereas Bob is 2 SDs above the mean. Thus, we conclude that Bob gave a better performance than John did.

The z-scores are needed to evaluate probabilities using normal distributions and will be discussed in the Appendix.

Z_α, known as the upper α percentile point of a standard normal distribution, is that z-score above which there are $100(\alpha)\%$ observations. Users of statistics must be able to evaluate z_α values for any given value α and this is described in the Appendix. Some of the commonly used z_α values are

$$z_{0.1} = 1.282, \qquad z_{0.05} = 1.645, \qquad z_{0.025} = 1.96,$$
$$z_{0.01} = 2.326, \qquad z_{0.005} = 2.576.$$

These z_α values indicate that there are 10% z-scores above 1.282, 5% z-scores above 1.645, 2.5% z-scores above 1.96, 1% z-scores above 2.326, and 0.5% z-scores above 2.576. A graphic representation of these z_α values are given in Figure 4.4.

Returning to the Example 4.8, it can be noted that because John's z-score was 1.4, between 5% and 10% of the students received higher scores than him. However, there are fewer than 2.5% of the students who received higher scores than Bob. Other examples using z_α values follow:

Example 4.9. (Did Bill Get a Lemon?) The length of time cars are used before requiring a major repair is normally distributed, say, with a mean $\mu = 3$ years and a SD $\sigma = 0.5$ years. Suppose Bill's car needed a major repair after being used for 1.25 years. The z-score of Bill's car usage is $(1.25 - 3)/0.5 = -3.5$. Thus Bill's car needed a major repair 3.5 standard deviations below the mean length needed for

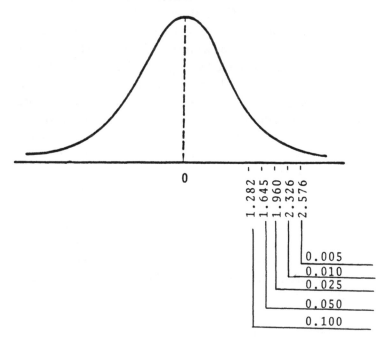

Figure 4.4 Upper percentile points of the standard normal variable.

major repairs. Many fewer than 0.5% of cars need repairs that early and Bill's car gave an exceptionally poor performance compared with the rest.

Example 4.10. (Revenue-Buster). Suppose the net revenue in the first week after a movie is released follows a normal distribution with mean $\mu = 5$ million dollars and a SD $\sigma = 1.7$ million dollars. Movie A produced a net revenue of 10.1 million dollars in the first week. The z-score of the revenue from movie A is $(10.1 - 5)/1.7 = 3$. The revenue from movie A is 3 SD above the mean revenue. Considerably fewer than 0.5% of movies collect a net revenue more than that of movie A. Statistically, movie A can be considered to have yielded a very high net revenue.

The normal distribution plays a key role in statistical inferences through the sampling distributions. Consider a population, whose observations are not necessarily normally distributed, with a mean μ_X and variance σ_X^2. Let several simple random samples with replacement of size n be taken from such a population. From each sample, a sample mean is calculated, each sample providing a sample mean. Thus, one obtains sample means $\bar{X}_1, \bar{X}_2, \bar{X}_3, \ldots$. The histogram of these sample means is that of an exactly normal distribution, when the population is normally distributed. Even when the population has an abnormal distribution, if the sample size n is large (say, $n > 30$), from a very important theorem in probability theory known as *central limit theorem,* the histogram of the sample means is shaped approximately as a normal distribution. Thus, in either event, the sample means are exactly or approximately normally distributed. The mean $\mu_{\bar{X}}$ and the standard deviation $\sigma_{\bar{X}}$ of the sample means are related to the population mean μ_X, population standard deviation σ_X, and sample size n given by the relations $\mu_{\bar{X}} = \mu_X$, $\sigma_{\bar{X}} = \sigma/\sqrt{n}$. $\sigma_{\bar{X}}$ is known as the standard error of the sample mean. Clearly, the standard error of the sample mean is smaller than the population standard deviation because of the division by $\sqrt{n}$. Inferences on the population mean will be based on the following normally distributed variable:

$$Z = \frac{\bar{X} - \mu_{\bar{X}}}{\sigma_{\bar{X}}} = \frac{\sqrt{n}(\bar{X} - \mu_x)}{\sigma_X}$$

The reader should not become confused by our using Z variable with different equations. The Z variable here still has $\mu_z = 0$, $\sigma^2_z = 1$. Z unifies all types of normal distributions. Earlier, Z was used to standardize the normally distributed random variable X, and, now, it is used to standardize the normally distributed random variable $\bar{X}$.

Often, the population standard deviation σ_X is unknown. In this event, the distribution of

$$\frac{(\bar{X} - \mu)\sqrt{n}}{s}$$

where the population is normally distributed and s is the sample standard deviation, is known as Student's T distribution with $n - 1$ degrees of freedom. The T distribution is similar to the normal distribution and is characterized by its degrees of freedom (see Figure 4.5). The degrees of freedom of the T distribution will be different from problem to problem. The degrees of freedom will be abbreviated by df and will be symbolically denoted by ν. The upper α percentile points of a T distribution with ν df will be written $t_{\alpha,\nu}$. A T distribution with degrees of freedom greater than or equal to 30, approximates a normal distribution.

The sample proportions $\hat{p}$s for large sample sizes (preferably $n \geq 100$; some people consider that n sufficient so long as $np > 5$) will also

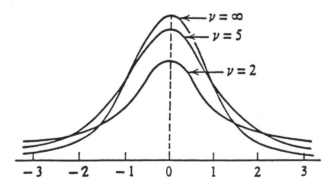

Figure 4.5   t-distribution curves for 2, 5 and ∞ d.f.

be normally distributed with mean $\mu_{\hat{p}} = p$ and standard error $\sigma_{\hat{p}} = \sqrt{p(1 - p)/n}$. Thus

$$Z = \frac{\hat{p} - p}{\sqrt{p(1 - p)/n}}$$

also follows a standard normal distribution.

The sampling distributions and their upper percentile points will be used in drawing required inferences concerning population means and proportions. The philosophy and the interpretation of these inferences will be discussed in the text. The mathematical, computational, and Minitab details involved in such inferences will be appended.

ESTIMATION

The simplest inference one can draw on a parameter is to estimate it based on the sample data. Consider the situation of an investigator interested in estimating the average family income in a community based on a sample survey. Without any guidelines for the functional form of the statistics used to estimate the average family income, the investigator may be biased in choosing the estimate and may use personal judgment, in estimating this value. Instead, if it is decided to use the sample mean as the estimator before the survey is made and the actual data collected, the investigator has no free choice of his/her own and is guided toward calculating the sample mean, which will estimate the average family income. In this connection the functional form of the statistic used to estimate a parameter is called an *estimator*, and the numerical value of the estimator based on the sample data is called the *estimate*. The sample mean and median are estimators of μ; s^2 is an estimator of σ^2; $\hat{p}$ is an estimator of p, etc. 15,500 dollars is the estimate of μ, the average family income in a community. 23 mpg is the estimate of μ the average fuel consumption of brand X cars. The number 0.3 is the estimate of p, the proportion of people who like Coca-Cola.

There are several possible estimators to estimate a parameter of interest. Of these, one should select the one with some desirable properties. There are several desirable properties listed in the literature. Two of the simple properties will be discussed here.

Given the estimator, it is conceivable to think that the estimates will vary from sample to sample because the observations of one sample need not be the same observation of another sample. Some estimates will be higher and some will be lower than the parameter. Very rarely does one obtain an estimate equal to the parameter that it is estimating. Statistics is a science of averages. It is one of the desirable properties of the estimator that, on an average, it equals the parameter it is estimating, and such an estimator is called an *unbiased estimator*. The mean of the sampling distribution of an unbiased estimator is the parameter that is estimated by it.

It is likely that several people or organizations conduct similar polls. The estimates of all of these polls will not, in general, be the same. If these estimates differ considerably, the validity of the scientific method of the surveys is questionable. Thus, it is appropriate to choose that estimator whose estimates based on several samples are closely clustered. In other words, one likes to choose that estimator whose standard error in the sampling distribution is the smallest. Such estimators are called *best estimators*. The estimates obtained from different samples for a best estimator will be closely clustered compared with the estimates obtained for any other estimate.

Estimators which are both best and unbiased are called the *best unbiased estimators*. The sample mean $\bar{X}$ is the best unbiased estimator of the population mean μ. The sample variance s^2 is the best unbiased estimator of the population variance σ^2. The sample proportion $\hat{p}$ is the best unbiased estimator of the population proportion p. Thus the sample mean is generally used to estimate the population mean and the sample proportion is used to estimate the population proportion.

Even when the best unbiased estimator is used, the estimate from a given sample need not necessarily be identical with the parameter it is estimating, because it has some standard error associated with it. Thus one may consider the best unbiased estimator, with some margin of error to estimate the parameter. Because events with a miniscule probability are still possible, it is still possible that the parameter estimated from an unbiased estimator with a margin of error may well be beyond the specified interval. If one wants to be absolutely certain of capturing the parameter within the interval, the only way of doing so is to specify the interval to be all possible values of the parameter. If a statistician estimates the average family income in a community as being any dollar amount, what good is that estimate? If meaningful and useful estimates must be obtained, it is absolutely necessary that the margin of error be small, and in that case a possibility exists of missing the parameter in the specified interval based on the data. The interval used to estimate the parameter is called a *confidence interval*. The percentage of samples for which the constructed confidence intervals cover the true parameter value is called the confidence coefficient. Procedures of calculating confidence intervals are given in the Appendix. In every problem there is an appropriate sample size to estimate the parameter with a given margin of error and a specified confidence coefficient. By taking a sample of size less than the required size, the margin of error will increase and samples of larger size are expensive. Appropriate sample size determinations are also given in the Appendix. Some examples of confidence intervals follow.

Example 4.11. (Average Family Income in a Community). An investigator polled a random sample of 100 resident families and found the sample mean $\bar{X}$ of the family incomes to be 15,000 dollars with a standard deviation of 5,000 dollars. Using the T-distribution with 99 degrees of freedom (approximately, normal distribution), a 95% confidence interval estimating μ, the population average family income is $15,000 \pm $980 (see Appendix). From this data one believes that μ is between $15,000 - $980 = $14,020, and $15,000 + $980 = $15,980. Only 95% of the samples have the property that the interval so constructed captures the real μ. The present sample may be one of the 95% such samples, in which case the calculated estimate is accurate. It is also possible that the current sample may be one of the 5% of the samples providing inaccurate estimates, in which case the estimate is wrong. Since people are usually optimistic, the investigator accepts the estimate of μ to be between $14,020 to $15,980 entertaining the possibility that he or she could be wrong. A 90% confidence interval for μ is 15,000 $\pm$ 822.50 dollars. A 99% confidence interval for μ is 15,000 $\pm$ 1,288 dollars.

This example indicates that the margin of error increases as one reduces the chance of being wrong.

Example 4.12. (Percentage of People Using Public Transportation). A poll of 400 people was conducted to determine the proportion p of people using public transport for commuting to be 0.3. A 95%

confidence interval for the proportion p of the population using public transport is 0.3 ± 0.04 (see Appendix). Thus the proportion p is between $0.3 - 0.04 = 0.26$ (26%) to $0.3 + .04 = .34$ (or 34%). The investigator believes p to be between 26% to 34% hoping that the sample used to calculate the estimate could be one of the 95% of the samples providing the correct estimate.

Example 4.13.   (Average Fuel Consumption). The average fuel consumption of brand X cars with given optional equipment was determined to be 22 mpg with a standard deviation of 2 mpg based on a sample of 36 cars. A confidence interval on the population average fuel consumption μ is 22 mpg $\pm$ 0.7 mpg (see Appendix). Thus μ is between 22 mpg $-$ 0.7 mpg $= 21.3$ mpg and 22 mpg $+$ 0.7 mpg $=$ 22.7 mpg. The investigator believes μ to be between 21.3 mpg and 22.7 mpg entertaining the possibility that the sample could be one of the 5% of the samples yielding a wrong estimate.

Example 4.14.   (Percentage of Defective Items). One hundred items randomly selected from a manufacturing process were quality tested and 3 items were found to be nonconforming to the specifications. A 95% confidence interval for the proportion of defectives, p, in that manufacturing process is 0.03 ± 0.03 (see Appendix). The proportion of nonconforming items in that establishment are expected to be between $0.03 - 0.03 = 0$ (or 0%) to $0.03 + 0.03 = 0.06$ (or 6%).

PHILOSOPHY OF TESTS OF STATISTICAL HYPOTHESES

The main ingredients of a statistical test are

1.   Null and alternative hypotheses (written as H_0 and H_A, respectively)
2.   Test Statistic
3.   Level of significance
4.   Decision rule
5.   Conclusion

Sometimes (4) and (5) are substituted by (3') p-value (or, observed significance level). Statistical tests using the ideas (1) through (5) are known in the literature as tests of hypotheses, and tests of significance are those using (3′) instead of (3) and (4). In this text, p-values will be considered with each test statistic. In this section these ideas will be discussed and will be illustrated with examples in the next section.

People generally formulate hypotheses and try to verify them. It is surprising, but easy, to note that no hypothesis can ever be verified. If a New Yorker is not mugged for 10 years, there is no guarantee that he or she will not be mugged in the eleventh year. A couple happily married for twenty years may suddenly develop differences and sepa-rate in the twenty-first. A volcano subdued for thousands of years can suddenly erupt. It can thus be seen that the hypothesis that a New Yorker is never mugged cannot be established. It can be disproved with a single encounter. Similarly, the couple who is happily married is a hypothesis that cannot be established, but that may be disproved. That the volcano is subdued and will never erupt also cannot be estab-lished.

In view of the discussion that no hypothesis can be proved, statisti-cians to establish claims use their complements as null hypotheses and try to find evidence in data to reject the formulated null hypothesis. Null hypotheses are denoted by H_0. By rejecting H_0, the alternate hypothesis, denoted by H_A, will be accepted. In other words, data must show strong evidence to favor H_A, whereas H_0 is retained if no strong evidence in favor of H_A is observed in data. In business prob-lems, there are consumers' and manufacturers' claims. Manufacturers' claims are weaker because, normally, they do not demonstrate their claims but leave the task of disproving their claims to others. Thus, manufacturers' claims are usually taken as H_0. When a consumer or consumer protection agency makes a claim, they must strongly support their claims, and thus, their claims are taken as H_A. H_0 and H_A are compliments of each other in the problems discussed in this book. However, in general they need not be complement to each other.

Three types of settings for null and alternative hypotheses arise in testing a single parameter. Suppose, a consumer protection agency wants to establish that the average fuel consumption, μ, of brand X cars is less than 16 mpg. The H_0 and H_A in their problem would be

$$H_0: \mu \geq 16 \text{ mpg}, \qquad H_A: \mu < 16 \text{ mpg}$$

because stronger evidence from the data is needed to establish that $\mu < 16$ mpg than to establish $\mu \geq 16$ mpg. In other situation, a chain store is considering opening a new branch in a community. The venture will be profitable if the average family income, μ, in that community is greater than 20,000 dollars. Here the store is interested to record a strong evidence from data that $\mu > 20,000$ dollars. Thus

$$H_0: \mu \leq \$20,000, \qquad H_A: \mu > \$20,000$$

A television network conducted a survey last year and found that 30% of the audience watched their new show, and is interested in knowing if this proportion, p, has changed this year. Here

$$H_0: p = 0.3, \qquad H_A: p \neq 0.3$$

In general, when testing the mean, μ, the three cases of formulating H_0 and H_A are

1. $H_0: \mu \geq \mu_o, H_A: \mu < \mu_o$
2. $H_0: \mu \leq \mu_o, H_A: \mu > \mu_o$
3. $H_0: \mu = \mu_o, H_A: \mu \neq \mu_o$

where μ_o is a numerical value based on the problem. Similarly, the three types of H_0 and H_A for testing proportions, p, are

1. $H_0: p \geq p_0, H_A: p < p_0$
2. $H_0: p \leq p_0, H_A: p > p_0$
3. $H_0: p = p_0, H_A: p \neq p_0$

where p_0 is a numerical value based on the problem. Depending on the specific objectives and goals of a study one and only one of the three types of formulations of H_0 and H_A will be selected and the conclusions will be drawn accordingly. It should be noted that the equal sign is always included in the statement of H_0. The null and alternative hypotheses of cases (1) and (2) are known as one-sided (or one-tailed) tests, and case (3) is known as a two-sided (or two-tailed) test.

After formulating the null and alternative hypotheses, one computes the test statistic, which measures the difference between the data and what is expected based on the null hypothesis. The test statistic is a random variable, while the computed quantity is a numerical value assumed by the random variable. Some show this difference by using different symbols. Such differences will not be shown here. In problems related to testing population means and proportions, the standard normal variable, Z, or the T-variable usually are the test statistics (see Appendix). The observed significance level (or p-value) is the probability of getting a test statistic as extreme as, or more extreme than the observed one, under the premise that H_0 is true. If this p-value is small, it is improbable to get the observed or more extreme differences for the data when H_0 is true, and H_0 is unlikely to be true. If the p-value is small, H_0 is rejected and H_A is accepted. If the p-value is not that small, H_0 is retained. Retention of H_0 does not mean that H_0 is accepted; it only implies that there is no evidence from the data to reject H_0. The smallness of the p-value is determined by the level of significance, α, which will be introduced now.

Consider a judiciary system. The motto of the judiciary system is that no one is guilty unless proved otherwise. When an accused who is charged with murder comes for trial, the judge will not formulate the null hypotheses that the accused is guilty and seek evidence to establish his innocence. In fact,

H_0: accused is innocent, H_A: accused is guilty.

The judge starts with the premise that the accused is innocent and looks for evidence, beyond a reasonable doubt, that the accused is guilty. In the process, several guilty people may escape sentence but occasionally an innocent person will be pronounced guilty. The accused may, in fact, be guilty or innocent and he or she alone knows the truth. Based on the evidence presented by prosecution and defense, the jury reaches a verdict and the judge may rule the accused as guilty or innocent. There are four possible situations now and they can be represented in the following format.

		Accused is actually	
		Guilty	Innocent
Judge rules the	Guilty	CORRECT	ERROR (Type I)
accused to be	Innocent	ERROR (Type II)	CORRECT

If a guilty person is found guilty, or an innocent person is found innocent, justice is done and the decision is correct. If a guilty person is released or an innocent person is sentenced, injustice is done and an error is committed in the decision. These two errors are distinguished by labeling them as Type I and Type II errors. If a true H_0 is rejected by the data, a Type I error is committed. Here, sentencing an innocent person is a Type I error. If a false H_0 is retained by the data, a Type II error is committed. Releasing a guilty person is a Type II error in this framework. The consequences and severity of these two errors will be different from problem to problem. As noted earlier, a Type I error in this setting is more severe than a Type II error.

Ideally, one likes to make the decisions without committing either of the two errors. However, if one type of error is avoided, the chances of committing the other type of error increase. The judge can completely eliminate Type I error by ruling every accused "not guilty." Then all the guilty criminals will also be ruled innocent and will be set free. Thus more Type II errors are committed by adopting such a practice. On the contrary, by ruling every accused to be guilty, the judge can eliminate committing a Type II error. Then more innocent people will be punished and more Type I errors are committed. Thus it is impossible to devise a decision process free from either errors.

A reasonable approach to handling these is to control the chance of committing a Type I error and that is called the *level of significance,* denoted by α. If the judge uses $\alpha = 0.01$, it is understood that he is ruling in such a fashion that not more than $(0.01)(100) = 1\%$ of innocent people are ruled "guilty." If an H_0 is rejected in favor of H_A at an α level of significance, it is understood that if similar data are collected and analyzed several times, $100\,\alpha\%$ times H_0 will be rejected when it is true. The sample at hand may be one of those odd $100\,\alpha\%$ samples in which H_0 is rejected. In a problem, if a Type I error is serious and one rarely wishes to commit such an error, α will be taken as small. If a Type II error is serious and one likes to minimize that error, α will be taken as large. By having α small (or large), fewer (or many) rejections of H_0 will be achieved in the decisions. Usually α will be taken as 0.05 or 0.01. If an H_0 is rejected using $\alpha = 0.05$, the test statistic is said to be significant. If an H_0 is rejected using $\alpha = 0.01$, the test statistic is said to be highly significant. The test statistic is not significant, if H_0 is not rejected at the 0.05 level.

If the p value of a test statistic is less than α, H_0 is rejected at a level of significance of α. Suppose for a test statistic, p-value is between 0.05 and 0.025. At an α of 0.01, H_0 is retained by that test statistic; while with $\alpha = 0.05$, H_0 is rejected. Thus the retention and rejection of H_0 completely depends on the chosen level of significance. However, it is again emphasized that smaller the p-value, the stronger is the evidence to reject H_0 in favor of H_A. In this book α is taken as 0.05 unless otherwise specified.

EXAMPLES OF TESTS OF HYPOTHESES

Several illustrations will be provided discussing all facets of statistical tests for testing population means and proportions.

Example 4.15. (Weight Gain of Babies). Suppose a pediatrician believes that the weight gain of babies during the first month is 3 lb. To test this claim a random sample of 50 babies were selected and the data on the first month weight gain yielded a sample mean, $\bar{X}$, of 3.1 lb, with a standard deviation of 0.4 lb. If μ is the average weight gain during the first month of all babies, the null and alternative hypotheses are

$$H_0: \mu = 3 \text{ lb}, \qquad H_A: \mu \neq 3 \text{ lb}$$

The weight gain data can be reasonably assumed to have a normal distribution and the population standard deviation σ is unknown. Hence, the test statistic is T with $50 - 1 = 49$ degrees of freedom. The calculated test statistic and p value for this data are (see Appendix for calculations),

$$T = 1.77, \qquad p\text{-value} = 0.0768$$

Because a T-distribution with more than 29 degrees of freedom is approximately a standard normal distribution, some statisticians report Z in place of T as the test statistic. If H_0 is true, there is a $(.0768)100 = 7.68\%$ chance of getting evidence against the null hypothesis as strong as the evidence at hand—or stronger. This probability is not small and thus there is no evidence against H_0. Thus the pediatrician's claim is retained. Since the p value is larger than 0.05, the test statistic is not significant.

Example 4.16. (Time Study). The average length of time to register for courses in a semester in a certain university has been 80 min.

A new registration procedure designed to shorten the registration time is being tried. In a random sample, 25 students were observed during registration; that data gave $\bar{X} = 55$ min and $s = 10$ min. The new procedure is useful, provided the average length of time for registration, μ, is less than 80 min. Thus H_0 and H_A for the problem at hand are

$$H_0: \mu \geq 80 \text{ min}, \qquad H_A: \mu < 80 \text{ min}$$

The length of time for registration can be considered to follow approximately a normal distribution, and the population standard deviation, σ, is unknown. The test statistic for this problem is T with $25 - 1 = 24$ df. The calculated T statistic and the p value (see Appendix) are

$$T = -12.5, \qquad p \text{ value} < 0.005$$

If H_0 is true, there is less than $(0.005)100 = 0.5\%$ chance of obtaining evidence against the null hypothesis that is as strong as the evidence at hand—or stronger. Thus, if H_0 is true, it is highly improbable that the observed or extreme departure from H_0 can be obtained from the data. Because such a departure is noticed in this data, H_0 is rejected and it is concluded that the new procedure reduced the average registration time to less than 80 min. Since p value is smaller than 0.01, the test statistic is highly significant.

Example 4.17. (Tar Content in Cigarettes). A cigarette manufacturer claims that the tar content does not exceed 17.5 mg. To test this claim, a random sample of nine cigarettes were tested, and it was found that $\bar{X} = 18.5$ mg., $s = 2.5$ mg. Because the manufacturer's claim is a weaker one, if μ is the average tar content in cigarettes, H_0 and H_A are

$$H_0: \mu \leq 17.5 \text{ mg.}, \qquad H_A: \mu > 17.5 \text{ mg}$$

The tar content in cigarettes approximately follows a normal distribution and the population standard deviation, σ, is unknown. Thus, the test statistic will be T with 8 degrees of freedom. The calculated test statistic T and the p value (see Appendix) are

$$T = 1.2, \qquad p \text{ value} > 0.1$$

If H_0 is true, there is more than $0.1(100) = 10\%$ chance of getting evidence against the null hypothesis as strong as the evidence at hand— or stronger. Such a situation is probable and there is nothing against the null hypothesis. H_0 is thus retained and the manufacturer's claim cannot be refuted. The p value is greater than 0.05 and the test statistic is not significant.

Example 4.18. (Percentage Defectives). A manufacturing company usually produces 5% nonconforming items. Some new quality control measures were instituted by the company to improve the quality. Afterwards, a random sample of 400 items were examined and 12 items are found defective. The purpose of the study is to minimize the percentage of defectives. If p is the proportion of defective items, the null and alternative hypotheses are

$$H_0: p \geq 0.05, \qquad H_A: p < 0.05$$

When testing for proportions the test statistic is the standard normal variable Z when the sample size n is large (preferably 100 or above). The calculated test statistic Z and the p value are (see Appendix)

$$Z = -1.84, \qquad p \text{ value} = 0.0329$$

If H_0 is true, there is a $0.0329(100) = 3.29\%$ chance of getting evidence against the null hypothesis as strong as the evidence at hand—or stronger. Such evidence is improbable and yet is there. Hence H_0 is rejected and it was concluded that the new measures decreased the percentage of defectives. The p value is between 0.05 and 0.01 and the test statistic, Z, is significant.

Example 4.19. (Automobile Mileage). It is claimed that Americans, on an average, drive more than 12,000 miles in each car they own. To test this claim, in a random sample, 100 cars were checked and the data yielded $\bar{X} = 15,000$ miles, $s = 2,500$ miles. Since the claim is to establish that μ, the average number of miles driven on a car exceeds 12,000 miles, the null and alternative hypotheses are

$$H_0: \mu \leq 12,000 \text{ miles}, \qquad H_A: \mu > 12,000 \text{ miles}$$

The mileage driven in the cars can be reasonably assumed to have a normal distribution, and the population standard deviation σ is unknown. The test statistic T with 99 degrees of freedom (can also be taken as approximately the standard normal variable Z). The calculated test statistic T and the p-value (see Appendix) are

$$T = 12, \qquad p\text{-value nearly } 0$$

If H_0 is true, there is nearly a 0% chance of getting evidence against the null hypothesis as strong as the evidence at hand—or stronger, and such an evidence is improbable. Since the data provides such an evidence, H_0 is rejected and it is concluded that Americans, on an average, drive more than 12,000 miles. The p value is less than 0.01 and the test statistic here is highly significant.

Example 4.20. (Balanced Coin). To test whether or not a coin is unbiased, 400 tosses of the coin were made and head was observed 212 times. If the coin is balanced, heads are expected 50% of the time and hence,

$$H_0: p = 0.5, \qquad H_A: p \neq 0.5$$

where p is the proportion of times heads occurs. This is a problem on proportions and the sample size is large. The test statistic is the standard normal variable Z. The calculated Z and the p value (see Appendix) are

$$Z = 1.2, \qquad p \text{ value} = 0.2302$$

If H_0 is true, there is a $0.2302(100) = 23.02\%$ chance of getting evidence against the null hypothesis as strong as the evidence at hand—or stronger, and such evidence is probable. Thus, H_0 will be retained and the coin will be considered as balanced. The p value is greater than 0.05 and the test statistic is not significant.

The above examples illustrate the method of drawing conclusions given H_0, H_A, test statistic and the p value in any specific problem.

APPENDIX

Sample Space and Probability

An experiment, as noted in Chapter 2, is a process of generating data. If the outcome of the experiment is not predetermined, it is called a random experiment. In statistics, only random experiments are considered. A complete enumeration of all possible outcomes of a random experiment in the form of a set is called a *sample space* and will be designated by S. Each possible outcome, that is, each element of S, is called a *simple event*, or *elementary event*, or sample point. A random event (or simply an *event*) is a collection of simple events possessing specified properties. Events will usually be denoted by capital roman letters. Events are subsets of the sample space in the set theoretic terminology.

Given two random events A and B, the event A or B (also called the sum of the events A and B) denoted by $A \cup B$, is the random event consisting of simple events belonging to A or B or to both. Clearly $A \cup B$ is the union of the sets A and B. The event A and B (also called the intersection of the events A and B) denoted by $A \cap B$ or AB, is the random event consisting of simple events belonging to both A and B. Clearly AB is the intersection of the sets A and B. The event not-A (also called the complement of the event A), denoted by $\bar{A}$, is the random event consisting of all simple events of S not belonging to A. Clearly A is the complement of the set A with respect to the universal set S. The events A and B are called disjoint if there is no common simple event between A and B. Clearly A and B are disjoint events if and only if $AB = \phi$, where ϕ is the empty set. Disjoint events are also called *mutually exclusive events*.

The following examples illustrate the ideas of sample space, events, and the algebra of events.

Example 4A.1. Let a random experiment consist of tossing a fair coin. One may get a head (H) or a tail (T) as the only outcomes of such an experiment. Thus $S = \{H, T\}$. There are two simple events in this experiment.

Example 4A.2. Let a random experiment consist of tossing a fair coin twice. Here $S = \{HH, HT, TH, TT\}$ with four possible outcomes. In each simple event, the first letter denotes the outcome of the first toss and the second letter denotes the outcome of the second toss. Let A be the event such that the first toss is a head, B be the event such that the second toss is a head, and C be the event such that there is at least one head. Then $A = \{HH, HT\}$, $B = \{HH, TH\}$, $C = \{HH, HT, TH\}$. None of the 3 pairs (A,B), or (A,C), or (B,C) are mutually exclusive because $AB = \{HH\} \neq \phi$, $AC = A \neq \phi$, and $BC = B \neq \phi$. The event

A or *B* is clearly *C*. The event *A* and *B* is {*HH*}, namely, getting a head on both tosses. The event not-*C* is {*TT*} representing a tail on both tosses.

Example 4A.3.   Let a random experiment consist of throwing a pair of fair dice. In problems of dice, it is better to think that the dice are of different colors; say red and blue. There are 36 possible outcomes, and

$$S = \{11, 12, 13, 14, 15, 16,$$
$$21, 22, 23, 24, 25, 26,$$
$$31, 32, 33, 34, 35, 36,$$
$$41, 42, 43, 44, 45, 46,$$
$$51, 52, 53, 54, 55, 56,$$
$$61, 62, 63, 64, 65, 66\}$$

Every simple event is a pair of numbers, the first representing the outcome of the red die and the second representing the outcome of the blue die. Let *A* be the event denoting a double, *B* be the event denoting a sum of 7 and *C* denoting at least one 6. Clearly $A = \{11, 22, 33, 44, 55, 66\}$, $B = \{16, 25, 34, 43, 52, 61\}$, $C = \{16, 26, 36, 46, 56, 61, 62, 63, 64, 65, 66\}$. *A* and *B* are mutually exclusive because $AB = \phi$, while *A* and *C* are not mutually exclusive because $AC = \{66\} \neq \phi$. $A \cup B$ is the event of getting a double or seven and is {11, 22, 33, 44, 55, 66, 16, 25, 34, 43, 52, 61}. $\bar{C}$ is the event of not getting a 6 and consists of 25 simple events. *A* and *C* represents a double 6.

Example 4A.4.   In a random experiment related to the home runs hit by a batter who gets up four times in a game, the sample space is

$$S = \{HHHH, HHHN, HHNH, HHNN, HNHH, HNHN, HNNH,$$
$$HNNN, NHHH, NHHN, NHNH, NHNN, NNHH, NNHN,$$
$$NNNH, NNNN\}$$

where *H* denotes a home run and *N* a no home run and the order of the four letters in each simple event denotes the outcome of his performance for the four times that he bats in that order. Let *A* be the event that he hits at least one home run. Then $\bar{A} = \{NNNN\}$.

Given a sample space *S*, probabilities (p_i) are the weights attached to the simple events such that

$$0 \leq p_i \leq 1, \qquad i = 1,2, \ldots, n \qquad \qquad (4A.1)$$

$$p_1 + p_2 + \cdots + p_n = 1 \qquad \qquad (4A.2)$$

where p_i is the weight given to the ith simple event for $i = 1,2, \ldots, n$. These weights, in a way, express the degree of belief of the investigator in realizing the outcomes if the random experiment is performed. There are many ways of assigning these weights, called probabilities. Of these, the most commonly used weights are the natural assignment of probabilities in which each simple event is given the same probability satisfying (4A.2). If S has n simple events, by using natural assignment of probabilities each simple event is given a weight of $1/n$. When the experiment is completely random and there is no reason to believe that some simple events are more likely to occur than others, the natural assignment of probabilities can be used. In Examples 4A.1, 4A.2, and 4A.3, the natural assignment of probabilities is appropriate. In Example 4A.4, it is unlikely that the batter can make four home runs or four no home runs with equal chances and hence natural assignment of probabilities is invalid.

The probability of an event A, denoted by $P(A)$, is the sum of all the probabilities of the simple events included in A. If n_A is the number of simple events in A and n is the number of simple events in the sample space, under the natural assignment of probabilities

$$P(A) = \frac{n_A}{n} \qquad \qquad (4A.3)$$

It is to be noted that $P(A)$ is approximately the proportion of times A occurs if the random experiment is repeated a large number of times.

The following rules sometimes will be useful in evaluating probabilities:

$$P(\phi) = 0 \qquad \qquad (4A.4)$$

$$P(A \cup B) = P(A) + P(B) - P(AB) \qquad \qquad (4A.5)$$

$$\bar{P}(A) = 1 - P(A) \qquad \qquad (4A.6)$$

The following examples illustrate the probability calculations:

Example 4A.5. Consider the random experiment of tossing a coin twice, as in Example 4A.2. Here the natural assignment of probabilities is valid. The number of simple events in S is $n = 4$. Clearly $n_A = 2$, $n_B = 2$, $n_C = 3$ and hence

$$P(A) = \frac{n_A}{n} = \frac{2}{4} = \frac{1}{2}$$

$$P(B) = \frac{n_B}{n} = \frac{2}{4} = \frac{1}{2}$$

$$P(C) = \frac{n_C}{n} = \frac{3}{4}$$

Example 4A.6. Consider the random experiment of throwing a pair of dice, as in Example 4A.3. Here the natural assignment of probabilities is valid. Clearly $n = 36$, $n_A = 6$, $n_B = 6$, $n_C = 11$, and hence

$$P(A) = \frac{n_A}{n} = \frac{6}{36} = \frac{1}{6}$$

$$P(B) = \frac{n_B}{n} = \frac{6}{36} = \frac{1}{6}$$

$$P(C) = \frac{n_C}{n} = \frac{11}{36}$$

If D denotes the event that the sum is 11, the event 7 or 11 is $A \cup D$ and one verifies that $n_{A \cup D} = 8$. Hence

$$P(A \cup D) = \frac{8}{36} = \frac{2}{9}$$

Example 4A.7. Consider the random experiment of Example 4A.4 related to the home runs of a batter in four times at bat. The natural assignment of probabilities is not valid here. Let the 16 simple events described in S be respectively denoted by $e_1, e_2, \ldots, e_{16}$. Let the probability assignment for the simple events be

$$p_1 = 0.0001, \qquad p_2 = p_3 = p_5 = p_9 = 0.0009$$
$$p_4 = p_6 = p_7 = p_{10} = p_{11} = p_{13} = 0.0081$$
$$p_8 = p_{12} = p_{14} = p_{15} = 0.0729, \qquad p_{16} = 0.6561$$

Then

$$P(A) = 1 - P(\bar{A}) = 1 - 0.6561 = 0.3439.$$

Example 4A.8. Let $S = \{e_1, e_2, e_3, e_4, e_5\}$ and the probability assignment to the simple events be $p_1 = 0.1, p_2 = 0.2, p_3 = 0.3, p_4 = 0.25, p_5 = 0.15$. Let $A = \{e_1, e_5\}, B = \{e_2, e_3, e_5\}$, and $C = \{e_1, e_2\}$.
Then

$$P(A) = p_1 + p_5 = 0.1 + 0.15 = 0.25$$
$$P(B) = p_2 + p_3 + p_5 = 0.2 + 0.3 + 0.15 = 0.65$$
$$P(C) = p_1 + p_2 = 0.1 + 0.2 = 0.3$$

To find $P(A \cup B)$, one notes $A \cup B = \{e_1, e_2, e_3, e_5\}$, and hence

$$P(A \cup B) = p_1 + p_2 + p_3 + p_5$$
$$= 0.1 + 0.2 + 0.3 + 0.15 = 0.75$$

To find $P(BC)$, one notes that $BC = \{e_2\}$ and hence

$$P(BC) = p_2 = 0.2$$

Example 4A.9. Consider a random experiment of drawing one card from a standard deck of 52 cards. The sample space consists of 52 simple events corresponding to the outcomes of drawing any one of the 52 cards. The natural assignment of probabilities is valid here and $n = 52$. Let A be the event of drawing a picture card or a 10 (that is, Ace, King, Queen, Jack or Ten). Clearly $n_A = 20$ because there are four cards in each of the five denominations, and

$$P(A) = \frac{n_A}{n} = \frac{20}{52} = \frac{5}{13}$$

Two events A and B are said to be *independent* if the occurrence or nonoccurrence of one event does not affect the chances of the occurrence of the other event. A pair of not independent events are said to be *dependent*.

Example 4A.10. If a coin is tossed two times and if A and B, respectively, denote the events of getting a head in the first and second tosses, A and B are independent events. Getting or not getting a head the first time does not alter the chances of getting a head in the second toss.

Example 4A.11.   Let John and Bob be two students in the same section of a first statistics course. Let A be the event that John gets an A grade and B be the event that Bob gets an A grade in the course. Then A and B are supposed to be independent events, unless John and Bob cheat on the examination.

Example 4A.12.   In a production line, let A be the event that the first selected item for quality check is defective and B be the event that the second selected item is defective. Then A and B are independent unless there is a malfunction in the machinery.

When A and B are independent events,

$$P(AB) = P(A) \cdot P(B) \tag{4A.7}$$

Example 4A.13.   A baseball batter has a record of hitting one home run on an average for every 10 times at bat. Let A be the event that he hits a home run in his first time at bat and B be the event that he hits a home run in the second time at bat in a game. A and B can reasonably be taken to be independent because whether or not he hits a home run in the first instance should not affect the chances of his hitting a home run in the second time at bat. His making home runs in the two times at bat is the event AB and

$$P(AB) = P(A) \cdot P(B) = (0.1)(0.1) = 0.01.$$

Mean and Variance of a Discrete Random Variable

Consider the pdf of a discrete random variable X given by

X	x_1	x_2	$\cdots$	x_k
$p_X(x)$	p_1	p_2	$\cdots$	p_k

Then

$$E(X) = \mu_X = x_1 \cdot p_1 + x_2 \cdot p_2 + \cdots + x_n \cdot p_n \tag{4A.8}$$

$$\sigma_X^2 = (x_1^2 \cdot p_1 + x_2^2 \cdot p_2 + \cdots + x_n^2 \cdot p_n) - \mu_x^2 \tag{4A.9}$$

Example 4A.14. From the pdf of X given in Example 4.5.,

$$\mu_X = 2\left(\frac{1}{36}\right) + 3\left(\frac{2}{36}\right) + 4\left(\frac{3}{36}\right) + 5\left(\frac{4}{36}\right) + 6\left(\frac{5}{36}\right) + 7\left(\frac{6}{36}\right)$$

$$+ 8\left(\frac{5}{36}\right) + 9\left(\frac{4}{36}\right) + 10\left(\frac{3}{36}\right) + 11\left(\frac{2}{36}\right) + 12\left(\frac{1}{36}\right)$$

$$= 7.$$

Example 4A.15. From the pdf of X given in Example 4.6,

$$\mu_X = 0(0.99) + 10,000(0.01) = 100$$
$$\sigma_X^2 = 0^2(0.99) + 10,000^2(0.01) - 100^2$$
$$= 990,000$$

Binomial, Poisson, and Hypergeometric Distributions

Let a random experiment consist of n independent trials, where each trial has only two outcomes called success (S) and failure (F). Let p be the probability of success in each trial and let p be the same from trial to trial. Let $q = 1 - p$. Let X be the number of successes in the n trials of the random experiment. Then, for a given numerical value r in a problem,

$$P(X = r) = \binom{n}{r} p^r q^{n-r}, \quad r = 0, 1, \ldots, n \quad (4A.10)$$

where $\binom{n}{r}$ is the number of combinations of r items taken from n items given by the formula

$$\binom{n}{r} = \frac{n!}{(r!)((n-r)!)} = \frac{1 \cdot 2 \cdots n}{(1 \cdot 2 \cdots r)(1 \cdot 2 \cdots (n-r))} \quad (4A.11)$$

and 0! is conventionally taken as 1. The probability given in (4A.10) is the $(r + 1)$th term in the binomial expansion of $(q + p)^n$ and hence the probabilities given in (4A.10) are said to follow a binomial distribution. The following examples illustrate the use of the binomial distribution.

Example 4A.16.   Let a test consist of 20 true–false multiple-choice questions and let X be the number of correct answers marked by a student guessing each answer without knowing the correct solution. The 20 questions are 20 trials for the student and hence $n = 20$. For each question, the student can provide a correct answer (S) or a wrong answer (F). Answering a question right or wrong has no influence on answering other questions correctly and hence the trials are independent. Since each question has only 2 answers, with probability $1/2$, the student can answer a question correctly with a random guess and thus $p = 1/2 = 0.5$ and $q = 1 - 0.5 = 0.5$. If one is interested in finding $P(X = 20)$, using formula (4A.10) it can be evaluated and is

$$P(X = 20) = \binom{20}{20}(.5)^{20}(.5)^{20-20} = \frac{20!}{(20!)(0!)}(.5)^{20}(1)$$

$$= 0.00000095$$

To find $P(9 \le X \le 11)$, one observes that $9 \le X \le 11$ implies that $X = 9$, or $X = 10$, or $X = 11$, and hence,

$$P(9 \le X \le 11) = P(X = 9) + P(X = 10) + P(X = 11)$$

$$= \binom{20}{9}(.5)^9(.5)^{20-9} + \binom{20}{10}(.5)^{10}(.5)^{20-10}$$

$$+ \binom{20}{11}(.5)^{11}(.5)^{20-11}$$

$$= \frac{20!}{9!11!}(.5)^9(.5)^{11} + \frac{20!}{10!10!}(.5)^{10}(.5)^{10}$$

$$+ \frac{20!}{11!9!}(.5)^{11}(.5)^9 = 0.4966$$

The probability that the student answers at least one question correctly is $P(X \ge 1)$ and is given by the formula (4A.6). Noting that $X = 0$ is the complement of $X \ge 1$,

$$P(X \ge 1) = 1 - P(X = 0)$$

$$= 1 - \binom{20}{0}(.5)^0(.5)^{20-0}$$

$$= 1 - 0.00000095 = 0.999999$$

Example 4A.17. Suppose a baseball player has a batting average of .400 and one is interested in finding the probability that he gets four hits in four times at bat. Each time this player is up to bat, is a trial and hence $n = 4$. Each time at the plate, he either hits (S) or misses (F). Hitting or not hitting in a trial has no influence on hitting at the other attempts and hence the trials are independent. Since his batting average is .400, he has 0.4 probability of hitting each time he is batting. Thus $p = 0.4$ and $q = 1 - 0.4 = 0.6$. Let X be the number of hits he makes. Then

$$P(X = 4) = \binom{4}{4}(0.4)^4(0.6)^0 = \frac{4!}{4!0!}(0.4)^4 = 0.0256.$$

The probability interpretation here is that in $(0.0256)100 = 2.56\%$ of the games when he gets up four times, he gets four hits. To find the probability that he gets at least one hit, one argues the same way as before and gets

$$P(X \geq 1) = 1 - P(X = 0)$$

$$= 1 - \binom{4}{0}(0.4)^0(0.6)^{4-0}$$

$$= 1 - \frac{4!}{0!4!}(1)(0.6)^4 = 1 - 0.1296 = 0.8704.$$

Thus in $0.8704(100) = 87.04\%$ of the games when he bats four times, he makes at least one hit.

Example 4A.18. Let a 95% efficient insecticide be sprayed on ten insects. Each insect provides a trial and hence $n = 10$. When the insect is exposed to the insecticide, either it will die (S) or it will survive (F). The death or survival of one insect does not influence the death of other insects and hence the trials are independent. Since the insecticide is 95% efficient, the probability that an insect will be killed with that insecticide is 0.95. Hence $p = 0.95$ and $q = 1 - 0.95 = 0.05$. Let X be the number of insects killed and suppose one is interested in finding the probability that at least 9 insects are killed. The required probability is

$$P(X \geq 9) = P(X = 9) + P(X = 10)$$

$$= \binom{10}{9}(0.95)^9(0.05)^{10-9} + \binom{10}{10}(0.95)^{10}(0.05)^{10-10}$$

$$= \frac{10!}{9!1!}(0.95)^9(0.05)^1 + \frac{10!}{10!0!}(0.95)^{10}(0.05)^0$$

$$= 0.3151247 + 0.5987369 = 0.9139$$

Poisson distribution is the limiting distribution of a binomial distribution when n is very large (preferably $n > 1000$) and p is very small ($p < 0.01$) such that $np = \lambda$ is finite. Then, if X is the total number of successes

$$P(X = r) = \frac{e^{-\lambda}\lambda^r}{r!}, \qquad r = 0, 1, 2, \ldots \qquad (4A.12)$$

where $e = 2.71828. \ldots$ Since the probability p is very small here, Poisson distribution is also known as the distribution of rare events. In some cases n and p will not be given, but the average number of successes will be given and that number can be taken as λ. The following examples illustrate the use of Poisson distribution.

Example 4A.19. On an urban highway 20,000 cars travel during weekday peak hours. The probability that any car has an accident is $1/10,000$. Of the 20,000 cars, any car can be involved in an accident and thus $n = 20,000$. Let S be the car's involvement in an accident and F be the car's noninvolvement in an accident. If multiple car accidents are ignored, the accidents can be taken to be independent. Here n is very large and p is very small. Thus $\lambda = np = (20,000)(1/10,000) = 2$. Let X be the number of accidents. Then

$$P(X = 1) = e^{-2}\frac{2^1}{1!} = 0.2707,$$

$$P(X = 2) = e^{-2}\frac{2^2}{2!} = 0.2707,$$

$$P(X \geq 1) = 1 - P(X = 0) = 1 - e^{-2}\frac{2^0}{0!} = 1 - 0.1353353$$

$$= 0.8647$$

Example 4A.20. Let 5000 people be injected with a vaccine and the probability that an individual suffers from a bad reaction from the vaccine is 0.001. One can verify that the number of people suffering from a bad reaction, X, follows a Poisson distribution with $\lambda = np = (5{,}000)(0.001) = 5$. Then

$$P(X = 5) = e^{-5} \frac{5^5}{5!} = 0.1755,$$

$$P(X \geqslant 1) = 1 - P(X = 0) = 1 - e^{-5} \frac{5^0}{0!} = 1 - 0.0067379$$

$$= 0.9933$$

Example 4A.21. A certain area is hit on an average by four hurricanes a year. The number of hurricanes, X, hitting that area in a year has a Poisson distribution with $\lambda = 4$ because it is the distribution of rare events. Now

$$P(X = 0) = e^{-4} \frac{4^0}{0!} = 0.0183$$

implying that $(0.0183)100 = 1.83\%$ of the years that area is spared from hurricanes. Again

$$P(X = 10) = e^{-4} \frac{4^{10}}{10!} = 0.0053$$

implying that $(0.0053)(100) = 0.53\%$ of the years that area gets 10 hurricanes.

While the trials in a binomial distribution are independent, the trials in a hypergeometric distribution are dependent. Let N be the number of items in a population of which k items belong to a given category C. Let a random sample of n items be selected without replacement from such a population and let X be the number of items of category C in the sample. Then

$$P(X = r) = \frac{\binom{k}{r}\binom{N-k}{n-r}}{\binom{N}{n}}, \qquad r = 0, 1, \ldots, \min(n,k) \quad (4A.13)$$

The following examples illustrate the use of this distribution.

Example 4A.22. From a deck of 52 playing cards, five cards are randomly selected. To find the probability that there are three aces in the selected five cards, hypergeometric distribution will be used. Since five cards are selected from 52 cards, $N = 52$ and $n = $ five. Since there are 4 aces in the deck, $k = 4$. Let X be the number of selected aces. Then $P(X = 3)$ is required and using (4A.13) this can be calculated as

$$P(X = 3) = \frac{\binom{4}{3}\binom{48}{2}}{\binom{52}{5}} = \frac{94}{54145} = 0.0017$$

Example 4A.23. Suppose a shipment of 10 items was delivered to a buyer and the shipment has two bad items. The buyer quality tests 3 items randomly selected from the shipment and accepts the shipment if no defective is found. To find the probability of accepting the shipment, one uses hypergeometric distribution. Since 3 items are randomly selected from 10 items, $N = 10$, $n = 3$. Since the shipment has 2 bad items, $k = 2$. Let X be the number of bad items selected for test by the buyer. The shipment is accepted if and only if $X = 0$ and the probability that the shipment is accepted is

$$P(X = 0) = \frac{\binom{2}{0}\binom{8}{3}}{\binom{10}{3}} = \frac{7}{15} = 0.4667$$

Normal Distribution

The standard normal variable is used for tabulating the areas under the normal curve and these are used to determine the probabilities from a normal distribution. If a and b are any real numbers and if X follows a normal distribution, then

$$P(a < X < b) = P(a < X \le b) = P(a \le X < b)$$
$$= P(a \le X \le b)$$

the common probability being the area enclosed between a and b under the normal curve of the variable X and the X-axis. One translates the probability statement involving X into a probability statement involving the standard normal variable Z. The required areas and probabilities can be calculated using the Table 1 at the end of the book, where the areas are given to the left of the z values, that is, $P(Z \leq z)$ are given in that table. The z values form the row and column labelling and the middle portion of the table corresponds to the area to the left of the z value. The z value up to the tenths are taken as row labeling and the hundredths place is taken as column labeling. When $z = 1.53$, one enters the table at 1.5 row and column .03 and reads the body of the table entry .9370. Thus $P(Z \leq 1.53) = 0.9370$; clearly $P(Z > 1.53) = 1 - P(Z \leq 1.53) = 1 - 0.9370 = 0.0630$. Again, when $z = -1.96$, one enters the table at -1.9 row and column .06 and reads the entry .0250. Thus $P(Z \leq 1.96) = 0.0250$. One should develop a mastery of finding probabilities using a normal distribution and the following examples are aimed for this purpose.

Example 4A.24. $P(Z \leq 2.33) = 0.9901, P(1 < Z < 2) = P(Z < 2.00) - P(Z \leq 1.00) = 0.9772 - 0.8413 = 0.1359, P(-1.96 < Z < 1.96) = P(Z < 1.96) - P(Z < -1.96) = .9750 - .0250 = .95$, $P(Z > 1.64) = 1 - P(Z \leq 1.64) = 1 - 0.9495 = 0.0505$.

Example 4A.25. Let X be normally distributed with mean $\mu_X = 12$ and $\sigma_X = 2$. Then

$$P(10 < X < 13) = P\left(\frac{10 - 12}{2} < \frac{X - 12}{2} = Z < \frac{13 - 12}{2}\right)$$

$$= P(-1 < Z < 0.5)$$

$$= P(Z < 0.50) - P(Z \leq -1.00)$$

$$= 0.6915 - 0.1587 = 0.5328$$

$$P(6 < X < 18) = P\left(\frac{6 - 12}{2} < \frac{X - 12}{2} = Z < \frac{18 - 12}{2}\right)$$

$$= P(-3 < Z < 3)$$

$$= P(Z < 3.00) - P(Z \leq -3.00)$$

$$= 0.9987 - 0.0013 = 0.9974$$

$$P(X \leq 15) = P\left(\frac{X - 12}{2} = Z \leq \frac{15 - 12}{2}\right)$$

$$= P(Z \leq 1.50)$$

$$= 0.9332$$

Note that the z-scores of the numbers are formed and used in evaluating the probabilities.

Example 4A.26. Suppose test scores, X, in a first statistic course are normally distributed with $\mu_X = 72$, and $\sigma_X = 8$. If A grades are given for those students who get a score of 80 or above, the proportion of A grades in the section are

$$P(X \geq 80) = P\left(\frac{X - 72}{8} = Z \geq \frac{80 - 72}{8}\right)$$

$$= P(Z \geq 1.00) = 1 - 0.8413 = 0.1587$$

Thus $0.1587(100) = 15.87\%$ of the students in the class receive an A grade. If students who do not make a score of 50 points fail the course, the proportion of F grades in the course is

$$P(X < 50) = P\left(\frac{X - 72}{8} = Z < \frac{50 - 72}{8}\right)$$

$$= P(Z < -2.75) = 0.0030$$

The upper α percentile value of a standard normal variable Z is denoted by z_α and it satisfies

$$P(Z > z_\alpha) = \alpha \qquad\qquad (4A.14)$$

To obtain z_α one enters the body of the table given in Table 1 with probability $1 - \alpha$ and notes the bordering z value. For example, $z_{0.025}$ is the z-value associated with the probability $1 - 0.025 = 0.975$. One notes that 0.975 occurs in the row numbered 1.9 and column headed by .06. Thus $z_{0.025} = 1.96$. $z_{0.05}$ is the z-value associated with the probability $1 - 0.05 = 0.95$. There is no entry of 0.95 in the body of the table. However entries 0.9495 and 0.9505 are in the table with associated z values of 1.64 and 1.65. Since 0.95 is the average of 0.9495 and 0.9505, the associated z-value is the average of 1.64 and 1.65. Thus $z_{0.05} = 1.645$. When $1 - \alpha$ is not in the table, the value closest to $1 - \alpha$ can be entered in the table, to find z_α. Mathematically more matured readers can find such z_αs from interpolation and it will not be discussed here. Clearly $100\alpha\%$ standardized observations will be below $-z_\alpha$. $-Z_\alpha$ is called the lower α percentile value.

Example 4A.27. Suppose the life of washing machines, X, is normally distributed with $\mu_X = 5$ years and $\sigma_X = 1.5$ years. If the manufacturer is willing to replace only 1% of the machines in the warranty period, the warranty period can be determined using z_α values. Let the company give a warranty of x years. Then x is found such that $P(X \leq x) = 0.01$, because the company is willing to replace only 1% of the machines during the warranty period. Now

$$P(X \leq x) = 0.01$$

$$P\left(\frac{X - 5}{1.5} = Z \leq \frac{x - 5}{1.5}\right) = 0.01$$

$$P\left(Z \leq \frac{x - 5}{1.5}\right) = 0.01$$

It is known that 1% of the standardized observations are less than $-z_{0.01} = -2.326$. Hence

$$\frac{x - 5}{1.5} = -2.326$$

$$x = 5 - (1.5)(2.326) = 1.511$$

Thus the company can give a warranty for 1.5 years, if it is willing to replace only 1% of the machines because of failure, in the warranty period.

Confidence Intervals

Confidence intervals are usually calculated using the sampling distribution of the best, unbiased estimator of the parameter. It is known from mathematical statistics that $\mu_{\bar{x}} = \mu$, $\sigma_{\bar{x}} = \sigma/\sqrt{n}$. The idea is now to capture the parameter in the middle using random end points based on the upper and lower percentile values of the distribution of the standardized best, unbiased estimator. The upper percentile values z_α of a standard normal variable were introduced earlier. The upper α percentile values of a t distribution with ν degrees of freedom are denoted by $t_{\alpha,\nu}$. $t_{\alpha,\nu}$ values are given in Table 2 at the end of the book for $\nu = 1, 2, \ldots, 29, \infty$; and $\alpha = 0.1, 0.05, 0.025, 0.01, 0.005$. To find $t_{\alpha,\nu}$ one reads the row labeled ν and column labeled t_α. Thus

$t_{0.05,10}$ is obtained by reading line 10 and column $t_{.05}$ and is 1.812. Thus $t_{0.05,10} = 1.812$. Since the t distribution approximates the standard normal distribution for large v, one uses $t_{\alpha,\infty} = z_\alpha$.

In the context of this chapter, three types of confidence intervals are needed. They are

1. A confidence interval for μ with confidence coefficient $1 - \alpha$ using sample mean $\bar{X}$ based on a simple random sample of size n, when σ is known, is

$$\bar{X} \pm (z_{\alpha/2})(\sigma/\sqrt{n}) \qquad (4A.15)$$

Usually σ^2 will be unknown and this result is only of theoretical interest.

2. A confidence interval for μ with confidence coefficient $1 - \alpha$ using sample mean $\bar{X}$ and sample standard deviation s based on a simple random sample of size n, when σ is unknown, is

$$\bar{X} \pm (t_{\alpha/2,n-1})(s/\sqrt{n}) \qquad (4A.16)$$

When $n > 30$, since $t_{\alpha/2,v} = z_{\alpha/2}$, some people replace (4A.16) by

$$\bar{X} \pm (z_{\alpha/2})(s/\sqrt{n}) \qquad (4A.16')$$

3. A confidence interval for a population proportion p with confidence coefficient $1 - \alpha$ using sample proportion $\hat{p}$ based on a simple random sample of size n (preferably $n > 100$) is

$$\hat{p} \pm (z_{\alpha/2})(\sqrt{\hat{p} \cdot \hat{q}/n}) \qquad (4A.17)$$

where $\hat{q} = 1 - \hat{p}$.

The calculation of these confidence intervals are demonstrated in the following examples.

Example 4A.28. Let $\bar{X} = \$15,000$, $s = \$5000$ and $n = 100$. A confidence interval for the population mean μ with confidence coefficient 0.95 (also known as 95% confidence interval for μ) can be calculated from (4A.16). Here $1 - \alpha = 0.95$ and $\alpha = 0.05$. The required confidence interval is

$$\bar{X} \pm (t_{0.025,99})\frac{s}{\sqrt{n}} = 15,000 \pm (1.96)\frac{5000}{\sqrt{100}} = 15,000 \pm 980$$

A 90% confidence interval for μ corresponds to an $\alpha = 0.1$ and is

$$\bar{X} \pm (t_{0.05,99})\frac{s}{\sqrt{n}} = 15{,}000 \pm (1.645)\frac{5000}{\sqrt{100}} = 15{,}000 \pm 822.50$$

Example 4A.29. Let $\bar{X} = 22$, $s = 2$, $n = 36$. A 95% confidence interval on the population mean μ corresponds to $\alpha = 0.05$ given by (4A.16) and is

$$\bar{X} \pm (t_{0.025,35})\frac{s}{\sqrt{n}} = 22 \pm (1.96)\frac{2}{\sqrt{36}} = 22 \pm 0.7$$

Example 4A.30. Let $\hat{p} = 0.3$, and $n = 400$. A confidence interval for the population proportion p with confidence coefficient 0.95 (also known as 95% confidence interval on p) is given by (4A.17) with $\alpha = 0.05$. The required interval is

$$\hat{p} \pm (z_{0.025})(\sqrt{\hat{p}\hat{q}/n}) = 0.3 \pm (1.96)(\sqrt{(.3)(.7)/400})$$
$$= 0.3 \pm 0.04$$

Example 4A.31. Let $\hat{p} = 0.03$ and $n = 100$. A 95% confidence interval on the population proportion p is

$$\hat{p} \pm (z_{0.025})(\sqrt{\hat{p}\hat{q}/n}) = 0.03 \pm (1.96)\sqrt{(0.03)(0.97)/400}$$
$$= 0.03 \pm 0.03$$

Sample Size for Estimating Mean

From (4A.15), the margin of error in estimating μ when σ is known using a confidence interval with confidence coefficient $1 - \alpha$ is $z_{\alpha/2}\sigma/\sqrt{n}$ and if the investigator is willing to control this error at δ, the required sample size n, satisfies

$$n \geq \frac{\sigma^2 z_{\alpha/2}^2}{\delta^2} \tag{4A.18}$$

Since larger sample sizes are expensive, n will be the smallest integer satisfying (4A.18).

Example 4A.32. Suppose an investigator wants to estimate the average family income in a city to within 1000 dollars using a confidence interval with confidence coefficient of 0.95. Furthermore, suppose σ is known from the previous surveys to be 2500. Here, $\delta = 1000$, $1 - \alpha = 0.95$ and $\sigma = 2,500$. Noting $z_{0.025} = 1.96$, from equation (4A.18), one gets

$$n \geq \frac{(2500)^2(1.96)^2}{(1000)^2} = 24.01$$

Rounding to the next higher integer, the sample size required for the survey is 25.

Example 4A.33. Suppose one is interested in determining the average gas consumption of brand X cars with a given equipment option, to within 1 mpg using a confidence interval with a 0.99 confidence coefficient. Also, the cars are expected to deliver at least 18 mpg and at most 24 mpg. Since the gas consumption in cars can be reasonably considered to follow a normal distribution, one can use the range of gas mileage, $24 - 18 = 6$ mpg to be approximately equal to 6σ. Thus $\sigma = 1$. Furthermore $\delta = 1$, $1 - \alpha = 0.99$, and $z_{0.005} = 2.576$. From (4A.18)

$$n \geq \frac{(1)^2(2.576)^2}{(1)^2} = 6.6$$

The number of cars needed for such an investigation is 7.

Sample Size for Estimating Proportion

From (4A.17), the error in estimating p using a confidence interval with confidence coefficient $1 - \alpha$ is $z_{\alpha/2}\sqrt{\hat{p}\hat{q}/n}$. If this error is controlled at δ, then

$$n \geq \frac{\hat{p}\hat{q}z_{\alpha/2}^2}{\delta^2} \tag{4A.19}$$

The formula (4A.19) cannot be applied because $\hat{p}$ is unknown. However, one substitutes a guess value for $\hat{p}$ and determines n. If no guess value for $\hat{p}$ is available, $\hat{p}$ will be taken as 0.5 in (4A.19). The sample size based on $\hat{p} = 0.5$ will be larger than the sample size based on any other $\hat{p}$.

Example 4A.34. Suppose a political scientist wants to estimate the percentage of votes that a Presidential Candidate receives in a primary using a confidence interval with a confidence coefficient of 0.95 to within 3 percentage points. Here $\delta = 0.03$, $1 - \alpha = 0.95$. If the candidate is expected to poll 30% of the votes, $\hat{p}$ will be taken as 0.3 and from (4A.19)

$$n \geq \frac{(0.3)(1 - 0.3)(1.96)^2}{(0.03)^2} = 896.6$$

A sample of size 897 is needed to meet the specifications. However, if $\hat{p}$ is unknown, $\hat{p}$ will be taken as 0.5 and

$$n \geq \frac{(0.5)(1 - 0.5)(1.96)^2}{(0.03)^2} = 1067.1$$

Under these circumstances, a sample of size 1068 is needed.

An investigator may have to pay a price by taking a larger sample size for his inability to guess a $\hat{p}$ value.

Tests of Hypotheses

The test statistic for testing any hypothesis about the population mean μ specified by a numerical value μ_0 for a one sided or two sided test is

$$Z = \frac{\bar{X} - \mu_0}{\sigma/\sqrt{n}} \tag{4A.20}$$

when σ is known, and

$$T = \frac{\bar{X} - \mu_0}{s/\sqrt{n}} \tag{4A.21}$$

with $n - 1$ degrees of freedom when σ is unknown. The test statistic for testing any hypothesis about the population proportion p, specified by a numerical value p_0 for a one sided or two sided test is

$$Z = \frac{\hat{p} - p_0}{\sqrt{p_0 q_0/n}} \tag{4A.22}$$

where $q_0 = 1 - p_0$ and n is large, preferably $n \geq 100$. The p value is the tail area from the calculated test statistic, the tail being determined based on the alternative hypothesis. The following examples will illustrate the calculation of the test statistic and the p-value.

Example 4A.35.   Consider Example 4.15, in which $H_0: \mu = 3$, $H_A: \mu \neq 3$ was tested using the sample data with $n = 50$, $\bar{X} = 3.1$, $s = 0.4$. This is a problem of testing the mean, when σ is unknown, and the test statistic of formula (4A.21) is

$$T = \frac{\bar{X} - \mu_0}{s/\sqrt{n}} = \frac{3.1 - 3.0}{0.4/\sqrt{50}} = 1.77$$

with 49 degrees of freedom. This T variable is approximately a standard normal variable Z. Because the test is two-sided, one should consider both tail areas for the p value and

$$p \text{ value} = 2p(Z > 1.77) = 2(1 - 0.9616) = 0.0768$$

Example 4A.36.   Consider Example 4.16 in which $H_0: \mu \geq 80$, $H_A: \mu < 80$ was tested using the sample data with $n = 25$, $\bar{X} = 55$, $s = 10$. The test statistic is the T variable of (4A.21) and is

$$T = \frac{\bar{X} - \mu_0}{s/\sqrt{n}} = \frac{55 - 80}{10/\sqrt{25}} = -12.5$$

with 24 degrees of freedom. Because the alternative hypothesis is $\mu < 80$, the p value is the tail area to the left of -12.5 for a T distribution with 24 degrees of freedom. In Table 2, there is an area of 0.005 to the right of 2.797 for a T distribution with 24 degrees of freedom. Thus to the right of 12.5 (or, equivalently to the left of -12.5) there is much less than a 0.005 area. Hence

$$p \text{ value} < 0.005$$

Example 4A.37.   Consider Example 4.17 in which $H_0: \mu \leq 17.5$, $H_A: \mu > 17.5$ was tested using the sample data with $n = 9$, $\bar{X} = 18.5$, $s = 2.5$. The test statistic is the T variable of (4A.21) and is

$$T = \frac{\bar{X} - \mu_0}{s/\sqrt{n}} = \frac{18.5 - 17.5}{2.5/\sqrt{9}} = 1.2$$

with 8 degrees of freedom. Because the alternative hypothesis is $\mu >$ 17.5, the p value is the tail area to the right of 1.2 for a T distribution with 8 degrees of freedom. In Table 2, there is an area of 0.1 to the right of 1.397 and hence there is an area of more than 0.1 to the right of 1.2 for a T distribution with 8 degrees of freedom. Hence

$$p\text{-value} > 0.1$$

Example 4A.38. Consider Example 4.18 in which $H_0: p \geq 0.05$, $H_A: p < 0.05$ was tested using the sample data with $n = 400$, $p = 0.03$. This is a problem of testing proportions using a large sample and the test statistic of (4A.22) is

$$Z = \frac{\hat{p} - p_0}{\sqrt{p_0 q_0/n}} = \frac{0.03 - 0.05}{\sqrt{(0.05)(0.95)/400}} = -1.84$$

Because the alternative hypothesis is $p < 0.05$, the p value is the area to the left of -1.84 for a standard normal distribution and, hence,

$$p \text{ value} = P(Z < -1.84) = 0.0329$$

Example 4A.39. Consider Example 4.19 in which $H_0: \mu \leq$ 12,000, $H_A: \mu > 12,000$ was tested using the sample data with $n = 100$, $\bar{X} = 15,000$, $s = 2500$. The test statistic is the T variable of (4A.21) and is

$$T = \frac{\bar{X} - \mu_0}{s/\sqrt{n}} = \frac{15,000 - 12,000}{2,500/\sqrt{100}} = 12$$

with 99 degrees of freedom, which can be considered as the approximate standard normal variable Z. Since the alternative hypothesis is $\mu > 12,000$, the p value is the area to the right of 12 under a standard normal distribution and hence

$$p\text{-value} = P(Z \geq 12) = \text{nearly } 0$$

Example 4A.40.   Consider Example 4.20 in which H_0: $p = 0.5$, H_A: $p \neq 0.5$ was tested using the sample data with $n = 400$, $\hat{p} = 0.53$. The test statistic Z of (4A.22) is

$$Z = \frac{\hat{p} - p_0}{\sqrt{p_0 q_0/n}} = \frac{0.53 - 0.50}{\sqrt{(0.5)(0.5)/400}} = 1.2$$

Because the test is two-sided, the p value is twice the tail area and hence

$$p \text{ value} = 2P(Z \geq 1.2) = 2(1 - 0.8849) = 0.2302$$

Minitab Computer Package

The confidence intervals and tests of hypothesis for μ can be done with a Minitab package when σ is known or unknown. A confidence interval for μ when σ is known is called a Z interval. It is called a T interval when σ is unknown. Similarly, a Z test or a T test give the test statistics and p value for testing the mean when σ is known or unknown.

After reading the data in C1, a 95 percent confidence interval on μ can be obtained, when $\sigma = .5$, say, using the Minitab command:

```
MTB > zinterval [95 percent confidence] sigma =
0.5, for c1
```

If σ is unknown, the following command provides a confidence interval for μ:

```
MTB > tinterval [95 percent confidence] for data in c1
```

The test statistic and p value of the hypothesis for H_0: $\mu = 10$ against the alternative H_A: $\mu \neq 10$, when σ is known, say $\sigma = 0.5$, can be obtained by using the command

```
MTB > ztest [of mu = 10] sigma = 0.5 on data in c1
```

For a one-sided alternative, the required command and subcommand are

```
MTB > ztest [of mu = 10] sigma = 0.5 on data in c1;
SUBC > alternative = -1 or +1.
```

In the subcommand type -1, if H_A is $\mu < 10$, and $+1$, if H_A is $\mu > 10$. When σ is unknown the T statistic and p value can be calculated from the command:

```
MTB > ttest [of mu = 10] on data in cl
```

Subcommand as before can be used for one-sided alternative hypotheses.

Example 4A.41. Suppose a sample of size 10 taken from a population with $\sigma = 0.5$ gave the following observations:

9.0, 10.4, 9.7, 9.6, 10.8, 10.1, 9.3, 10.6, 9.9, 10.3

The investigator is interested in getting a 95 percent confidence interval for μ and in finding the test statistic, Z, for testing H_0: $\mu = 10$, H_A: $\mu \neq 10$. The Minitab commands for inputting the data and executing the program, and computer output are as follows:

```
MTB>read cl
DATA>9.0
DATA>10.4
DATA>9.7
DATA>9.6
DATA>10.8
DATA>10.1
DATA>9.3
DATA>10.6
DATA>9.9
DATA>10.3
DATA>end
      10 ROWS READ
MTB>zinterval [95 percent confidence] sigma=0.5 for cl

THE ASSUMED SIGMA = 0.500

       N    MEAN    STDEV   SE MEAN   95.0 PERCENT C.I.
Cl    10   9.970   0.577     0.16    (9.66,   10.28)

MTB>ztest [of mu=10]sigma=0.5 on data in cl

TEST OF MU = 10.0 VS MU N.E. 10.0
```

THE ASSUMED SIGMA = 0.500

	N	MEAN	STDEV	SE MEAN	Z	P VALUE
C1	10	9.970	0.577	0.16	-0.19	0.85

If σ is unknown, a confidence interval could have been determined.

MTB>tinterval [95 percent confidence] for data in c1

	N	MEAN	STDEV	SE MEAN	95.0 PERCENT C.I.
C1	10	9.970	0.577	0.18	(9.56, 10.38)

If σ is unknown and if the investigator is interested to test H_0: $\mu \geq 10$, H_A: $\mu < 10$, the following procedure will be used.

MTB>ttest [of mu=10] on data in c1;
SUBC>alternative=-1.

TEST OF MU = 10.0 VS MU L.T. 10.0

	N	MEAN	STDEV	SE MEAN	T	P VALUE
C1	10	9.970	0.577	0.18	-0.16	0.44

EXERCISES

1. If a balanced coin is tossed three times, find the probability of getting (a) exactly one head, (b) exactly two heads, (c) at least one head.
2. Verify the pdf given in Example 4.5.
3. Are a pair of disjoint events always dependent?
4. From a bowl containing four red and two green balls, one ball was randomly selected. What is the probability that it is a red ball?
5. Find μ_X and σ_X^2 for the following pdf.

X	-2	0	1	2
$p_X(x)$	0.3	0.1	0.4	0.2

6. From a sample of wheat seed, 10% are nongerminating. If five seeds are sown, what is the probability that (a) none germinate, (b) all germinate, (c) at least one germinates?

7. It is believed that 30% of people like Coca-Cola. If a random sample of 20 people are asked their preference, what is the probability that (a) none like Coca-Cola, (b) at most five people like Coca-Cola, (c) at least ten people like Coca-Cola?

8. The average number of worms per square inch area is 10.0. What is the probability of finding (a) at least 1, (b) at most 2, (c) exactly 10 worms in a given square inch area?

9. A typist, on an average, commits one mistake per each typed page. What is the probability that on the next page she types there are (a) no mistakes, (b) five mistakes, (c) at least one mistake?

10. It is believed that 25 students in an elementary statistics class of 40 students do not like statistics. If a random sample of three students are selected from the class and are asked about their opinion of statistics, what is the probability that all of them dislike statistics?

11. The average seasonal rainfall in a city is 20 inches with a standard deviation of 4 inches. What is the probability that in a year the rainfall in that city will be (a) between 25 and 30 inches, (b) at most 15 inches, (c) at least 30 inches?

12. The finished inside diameter of a piston ring is normally distributed with a mean of 4 inches and a standard deviation of 0.01 inch. What proportion of rings will have inside diameter (a) exceeding 4.02 inches, (b) between 3.99 and 4.01 inches?

13. A random sample of 100 light bulbs manufactured by a company has an average life of 800 hours with a standard deviation of 40 hours. Find a 90% confidence interval for the population mean life of all bulbs produced by that company.

14. A random sample of 25 business executives has an average IQ of 110 points with a standard deviation of 5 points. Find a 95% confidence interval on the population mean IQ of all business executives.

15. The profit per car made by an used car dealer based on a random sample of 16 cars has an average of 500 dollars with a standard deviation of 100 dollars. Find a 95% confidence interval for the population mean profit per car of that dealer.

16. Determine a 95% confidence interval for the proportion of defective items in a manufacturing process when it is found that a sample size of 200 yields 10 defectives.

17. In a random sample of 200 students in an urban university, 150 students use public transportation. Find a 90% confidence interval for the proportion of students of that university using public transportation.

18. In a random sample of 1000 patients treated with a new drug, adverse reactions were observed in 10 patients. Compute a 99% confidence interval for the population proportion of patients who will have adverse reaction to that drug.

19. A feeding experiment conducted on 100 experimental animals showed an average weight increase of 10 lb, with a standard deviation of 2 lb. The manufacturer of the feed claims that his product produces a gain of at least 11 lb. Is there enough reason to reject the manufacturer's claim?

20. An insecticide killed 110 of 120 insects. Do you agree with the claim that the insecticide is 95% efficient?

21. The average income of a random sample of nine families is 20,000 dollars with a standard deviation of 3,000 dollars. Is there enough evidence to believe that population mean income of that community is greater than 18,000 dollars?

22. The average number of miles driven between oil changes based on a random sample of 25 drivers, is 8000 with a standard deviation of 750 miles. Does this data support the claim that the average mileage driven between oil changes is at most 7500 miles?

23. In a random sample of 100 Philadelphians, 70 people watched the Live Aid Concert. Is there enough evidence to believe that more than 60% of Philadelphians watched that concert?

24. In a series of experiments, 531 plants with green foliage and 469 with yellow foliage were counted. This was a back cross in which 50% of the plants are expected to have green foliage. Is the data compatible with the theory?

25. In a random sample of 1000 women aged 45 and over, 55 have breast cancer. Does this data support the claim that at most 5% women of that age group suffer from breast cancer?

Answers

(1a) 3/8; (1b) 3/8; (1c) 7/8; (3) yes; (4) 2/3; (5) $\mu = 0.2$, $\sigma^2 = 2.36$; (6a) 0.00001; (6b) 0.5905; (6c) 0.99999; (7a) 0.0008; (7b) 0.4163; (7c) 0.0479; (8a) 0.999955; (8b) 0.0028; (8c) 0.1251; (9a) 0.3679; (9b) 0.0031; (9c) 0.6321; (10) 0.2328; (11a) 0.0994; (11b) 0.1056; (11c) 0.0062; (12a) 0.0227; (12b) 0.6826; (13) 800 ± 6.58; (14) 110 ± 2.064; (15) 500 ± 53.28; (16) 0.05 ± 0.03; (17) 0.75 ± 0.05; (18) 0.01 ± 0.008; (19) $t = -5$ with 99 df, p value nearly zero; (20) $Z = -1.68$, p value $= 0.0465$; (21) $t = 2$ with 8 df, p value between 0.025 and 0.05; (22) $t = 3.33$ with 24 df, p value less than 0.005; (23) $Z = 2.04$, p value $= 0.0207$; (24) $Z = 1.96$, p value $= 0.05$; (25) $Z = 0.73$, p value $= 0.2327$.

5

Use of the Relationship Between Variables

From Metropolitan Life Tables (see *Time Magazine,* August 19, 1985, p. 57) one notes, for example, that men between the ages of 25 and 29 with 5 ft. 9 in. height can weigh anywhere between 139 and 175 lb. This implies that weight is a dependent variable on the height of the people. The average weight of men between the ages of 25 and 29 with 5 ft. 9 in. height is estimated at 157 lb. with a margin of error of 18 lb., and thus the weights of such people varies from $157 - 18 = 139$ lb. to $157 + 18 = 175$ lb. It is obvious that not all people of that group weigh within the specified limits. It is only a probabilistic statement and people weighing outside of those limits are considered too thin or too fat. Sometimes observations on several variables will be used to predict the observation for another variable. Problems of this type are solved by *regression methods,* and will be discussed in this chapter. The study of correlation between variables is interconnected to the regression analysis and also will be considered here.

142

SCATTER DIAGRAM

In the simplest example, when two variables X and Y are observed for each of n units, the type of functional form to predict Y for given X can be determined by examining the scatter diagram. The scatter diagram is a graph of n points corresponding to the units. Each unit is marked with its X variable observation as its x coordinate and the Y variable

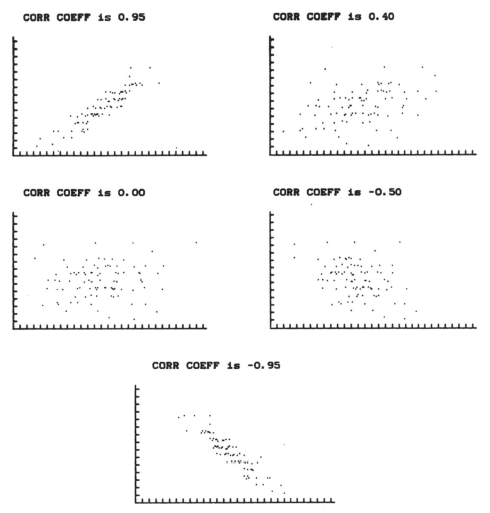

Figure 5.1   Scatter diagrams for selected correlation coefficients.

observation as its y coordinate. A measure of linear relationship be-tween the X and Y variables is the correlation coefficient. The popula-tion correlation coefficient is denoted by ρ and the sample correlation coefficient is denoted by r. The correlation coefficient always lies between -1 and 1, including the end values. If the correlation coeffi-cient is -1 or 1, there is a perfect linear relationship between X and Y and if the correlation coefficient is zero, there is no linear relationship between X and Y. In Figure 5.1, the scatter diagram for some selected correlation coefficients are indicated. If the correlation coefficient is positive, the Y values increase with increasing X values. If the correla-tion coefficient is negative, the Y values decrease as X values increase. The calculation of correlation coefficient is described in the Appendix.

The following examples illustrate the role of scatter diagrams in deciding the functional form of the equation to predict the dependent variable.

Example 5.1. (First and Second Statistics Courses). Stat 21 and Stat 22 are two statistics courses taken by all undergraduate students of

Table 5.1 Test Scores of 10 Students in Stat 21 and Stat 22

Student number	Test score in	
	Stat 21 (X)	Stat 22 (Y)
1	64	72
2	75	65
3	72	69
4	71	78
5	58	70
6	68	88
7	72	73
8	74	80
9	80	76
10	80	73

the School of Business Management at Temple University. The test scores obtained by 10 students in both courses are given in Table 5.1. To decide the functional form for predicting Y given X (i.e., to predict the test score in Stat 22 given the test score in Stat 21), a scatter diagram given in Figure 5.2 is plotted. From the scatter diagram given in Figure 5.2, it is clear that there is no relationship between X and Y, and the points are spread out over the entire plane. Thus given this data it is not possible to formulate a regression equation to predict Y. This is supported by the fact that the correlation coefficient between X and Y for the data of Table 5.1 is 0.019 (see Appendix).

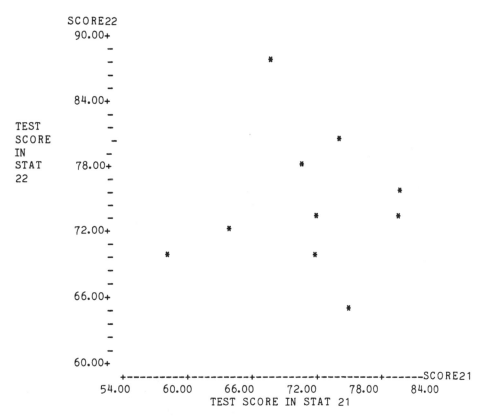

Figure 5.2   Scatter diagram for the data of Table 5.1.

Example 5.2. (Population and GNP). The mid-1976 population (*X*) and GNP per capita in dollars, (*Y*), for selected Western European countries are given in Table 5.2. To see the relationship between GNP per capita and population based on the available data in Table 5.2, one can form the scatter diagram given in Figure 5.3. In the scatter diagram shown in Figure 5.3, the presence of two points with almost identical

Table 5.2 Mid-1976 Population, (*X*), and GNP per Capita, (*Y*), in Dollars for Selected Western European Countries

Country	Population in millions (*X*)	GNP per capita (*Y*)
Austria	13.8	5415
Belgium	9.8	5829
Denmark	5.1	6461
Finland	4.7	4313
France	53.2	6312
Germany (Fed. Rep.)	61.6	6853
Greece	9.1	2128
Iceland	0.2	4455
Ireland	3.2	2536
Italy	56.2	3154
Luxembourg	0.4	5806
Malta	0.3	1300
Netherlands	13.8	5415
Norway	4.0	6449
Portugal	9.7	1239
Spain	35.8	2212
Sweden	8.2	7900
Switzerland	6.4	7283
United Kingdom	56.0	3844

Source: *Time*, the 1979 *Hammond Almanac*, p. 199. Reproduced with the permission of the Publisher.

coordinates is indicated by the numeral 2. The points also are widely spread out over the graph and there is no apparent relation between the two variables. One cannot hope to predict GNP per capita dollars based on the population of a country. Furthermore, the correlation coefficient between X and Y for this data is only 0.031.

   Example 5.3. (Age and Blood Pressure). It is well known that the blood pressure (BP) of people depends on age. The age (X) and systolic blood pressure (Y) for 11 males are given in Table 5.3. To decide the functional form to predict Y given X value, a scatter diagram is plotted in Figure 5.4. From the scatter diagram shown in Figure 5.4, it is apparent that the points are very close to a straight line with a

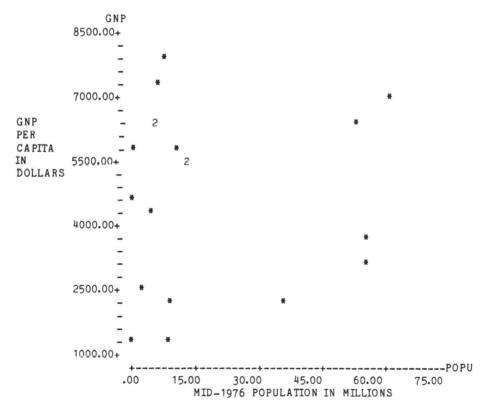

Figure 5.3   Scatter diagram for the data of Table 5.2.

Table 5.3 Artificial Data on Age, (X), and Systolic BP, (Y), for Men

Person number	Age (X)	BP (Y)
1	56	147
2	42	125
3	60	140
4	49	135
5	55	146
6	48	136
7	38	110
8	42	115
9	60	148
10	50	140
11	50	140

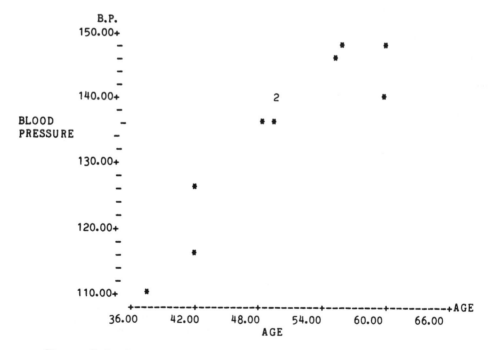

Figure 5.4 Scatter diagram for the data of Table 5.3.

positive slope. Furthermore, the correlation coefficient between X and Y is 0.901 (see Appendix), indicating the appropriateness of using a straight line equation to predict Y. The mathematical equation of a straight line is $Y = \beta_0 + \beta_1 X$. Since the observations on X and Y need not be exactly on a straight line, introducing an error term (e) to explain the departures of the points from the line, a model $Y = \beta_0 + \beta_1 X + e$ will be considered to predict Y given X. The coefficients β_0 and β_1 will be estimated from the data by b_0 and b_1 by the *method of least squares*, and Y will be predicted by $\hat{Y}$ given by the equation $\hat{Y} = b_0 + b_1 X$. The equation $\hat{Y} = b_0 + b_1 X$ will be called the sample linear regression equation of Y on X.

Example 5.4. (Growth of Bacteria). The bacteria (Y) were counted at different points of time (X) in a bacterial colony and data are given in Table 5.4. The scatter plot of the data is shown in Figure 5.5. From Figure 5.5 it appears that although a linear relationship $\hat{Y} = b_0 + b_1 X$ may reasonably predict Y, more appropriately, the curve is elbow-shaped and a growth curve of the type $\hat{Y} = b_0 b_1 X$, should be used to predict Y for given X. An equation $\hat{Y} = b_0 b_1^X$ is equivalent to $(ln\ \hat{Y}) = (ln\ b_0) + X(ln\ b_1)$ and thus, a linear regression will be used to predict $Z = ln\ \hat{Y}$ for a given X, and $\hat{Y}$ can then be calculated from the predicted $\hat{Z}$. It may be noted that $ln\ \hat{Y}$ is the natural logarithm of $\hat{Y}$. The correlation coefficient between X and Y is 0.954 indicating the adequacy of a straight line fit also. The fitting of both types of equations to this data will be discussed in the next section. The model used in this connection is $Y = \beta_0 \beta_1^X e$ and the error is multiplicative with the other terms.

Example 5.5. (Production and Profit). The production (X, in thousands of items) and the net profit per item (Y, in dollars) for eight small manufacturing companies are given in Table 5.5. The scatter plot for this data is shown in Figure 5.6. The correlation coefficient between X and Y is -0.233 and the points clearly do not form a straight line. Hence, a straight-line regression equation for predicting Y is out of the question. The points reasonably form a parabola and, hence, one can attempt a quadratic regression equation of the form $\hat{Y} = b_0 + b_1 X + b_2 X^2$, where b_0, b_1, and b_2 will be estimated from the data. The model used here is $Y = \beta_0 + \beta_1 X + \beta_2 X^2 + e$.

Table 5.4 Artificial Data on Bacteria
Count, (Y) at Different Time Periods, (X)

Time (hr) (X)	Number of bacteria (Y)
0	32
1	47
2	65
3	92
4	132
5	190
6	275

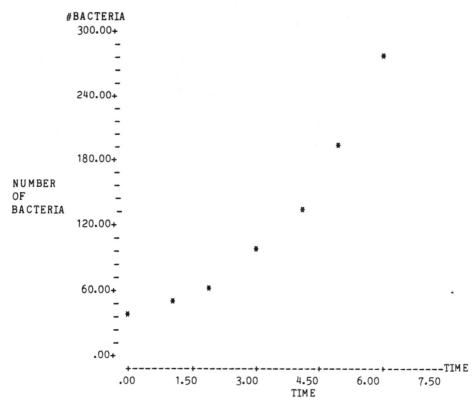

Figure 5.5 Scatter diagram for the data of Table 5.4.

Table 5.5 Artificial Data on Production, (X), and Net Profit, (Y) of 8 Companies

Company number	Production (1000 item) (X)	Net profit/item (in dollars) (Y)
1	5	10
2	7	12
3	8	13
4	10	15
5	12	14
6	13	12
7	15	11
8	20	10

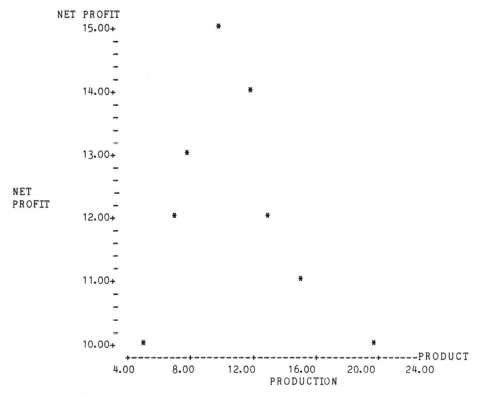

Figure 5.6 Scatter diagram for the data of Table 5.5.

LINEAR REGRESSION EQUATION

When the points in a scatter diagram almost cluster along a straight line with positive or negative slope, one predicts the dependent variable Y given the independent variable (also known as predictor variable) X, using an equation of the form $\hat{Y} = b_0 + b_1X$, where b_0 and b_1 are estimates of the model parameters β_0 and β_1 determined from data, and $\hat{Y}$ is the fitted value. The method of determining b_0 and b_1 is the *method of least squares*. Under the assumptions that Y values at each X value are normally distributed, with a common variance and that the mean values of Y for each X form a straight line, b_0 and b_1 will be determined from data. This situation can be sketched as in Figure 5.7. The mathematics of this procedure is appended to this chapter.

There are several computer packages available for regression problems. In this book, Minitab outputs will be discussed and the reader is expected to have gained working knowledge of Minitab through the Appendices of the last two chapters, or has access to get Minitab output. For making reasonable interpretations of the data and to do some diagnostic checks, the data on Y and X can be entered in C1 and C2 as described in the Appendix of Chapter 3. The columns can be named by the names of the variables of Y and X using the NAME

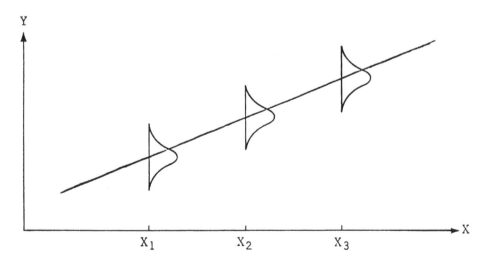

Figure 5.7 Assumptions of simple linear regression equation.

command. The name should not exceed six letters and should be a single word. One may use the following commands:

MTB > name cl is 'y'
MTB > name c2 is 'x'

The acceptance of a name command is prompted by the computer by MTB >. The computer outputs obtained after each of the following three Minitab commands excluding the name command are needed for making a reasonable statistical analysis of a problem. Print a command after the prompt MTB > appears on the screen.

MTB > regress cl (1) c2 (store studentized residuals in
      c3)
MTB > name c3 is 'tresd'
MTB > print c2, cl, c3
MTB > plot c3 vs c2

The Minitab run for the data of Table 5.3 is given now, where $\hat{Y}$ is written, instead of Y, in describing the regression equation.

THE REGRESSION EQUATION IS

$\hat{Y} = 56.2 + 1.57\ X$

COLUMN	COEFFICIENT	ST. DEV. OF COEF.	T-RATIO = COEF/SD
	56.20	12.71	4.42
X	1.5706	0.2518	6.24

S = 5.840

R-SQUARED = 81.2 PERCENT
R-SQUARED = 79.1 PERCENT, ADJUSTED FOR DF

ANALYSIS OF VARIANCE

DUE TO	DF	SS	MS=SS/DF
REGRESSION	1	1327.2	1327.2
RESIDUAL	9	307.0	34.1
TOTAL	10	1634.2	

X	Y	TRESD
56	147	0.53154
42	125	0.54662
60	140	-2.10065
49	135	0.33136
55	146	0.63040
48	136	0.79591
38	110	-1.25699
42	115	-1.37959
60	148	-0.48997
50	140	0.94686
50	140	0.94686

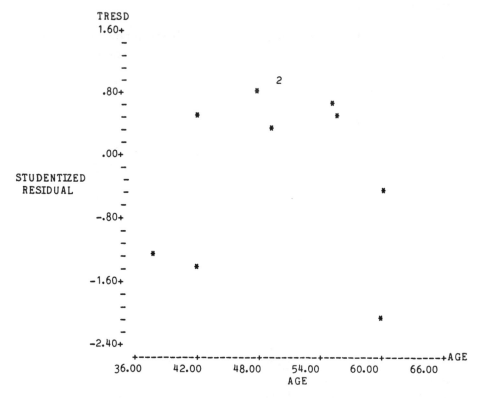

Figure 5.8   Residual plot.

Interpreting the Regression Equation

The equation to predict the average blood pressure for men of X years of age is $\hat{Y} = 56.2 + 1.57X$. When $X = 40$, $\hat{Y} = 56.2 + 1.57(40) = 119$. Each male aged 40 years need not have a BP measurement of 119. The average BP of all men aged 40 is estimated from the data as 119. This is also equivalent to saying that each man aged 40 years is expected to have a BP measurement of 119. The interval estimates of this quantity will be discussed later.

In the equation, 56.2 is called the *Y-intercept* or the constant term. It is the value of $\hat{Y}$ when $X = 0$. In most problems, it does not provide a meaningful quantity. In the present context, 56.2 has no meaningful interpretation, because men with $X = 0$ do not exist and their BPs cannot be measured. It helps to get a reasonable equation to fit the data. If the experimenter has reason to believe that $Y = 0$ when $X = 0$, that is, the regression line passes through the origin, the equation may be fitted without the constant term option using appropriate Minitab subcommand.

In the equation, 1.57 is the coefficient of the independent variable X and is called the slope or sample regression coefficient of Y on X. It was noted before that the sample regression coefficient b_1 estimates an analogous quantity β_1 in the model. In this case an estimate of the population regression coefficient β_1 is $b_1 = 1.57$.

The regression coefficient measures the change of Y values for a unit change in X values. It can also be interpreted as the rate of change of the Y values. For each year a person ages, the BP is expected to increase by 1.57 units. Negative regression coefficients will be interpreted as the decrease in Y values when X is increased by 1 unit.

The data are collected in this situation for the X variable ranging from 38 to 60. The use of the predicting equation can be made for X values within this range. By using the equation to predict Y for X values, for those values beyond the range of recorded X values, erroneous and inaccurate estimates will be obtained. The calculated equation in this problem is not recommended for predicting the BP of men aged 20. It is also not recommended for predicting the BP of men aged 75. It works reasonably well to predict the BP of men aged 38 to 60. The prediction is more accurate when the independent variable is $\bar{X}$ or close to $\bar{X}$. In this problem $\bar{X} = 50$. The equation most accurately estimates the average BP of men aged 50 as $56.2 + (1.57)(50) = 134.7$. It may be noted that $\hat{Y} = \bar{Y}$ when $X = \bar{X}$.

Inferences About the Population Regression Coefficient

We would like to know if the information provided by the independent variable is useful to predict the dependent variable. If the population regression coefficient, β_1, is zero, there is no change in Y values for changes in X values, and, in such cases, one predicts Y as $\bar{Y}$, irrespective of the X value. On the contrary, if $\beta_1 \neq 0$, the Y values change with changing X values, and the information of the X variable will be used to predict the Y values. Thus, one tests the null hypothesis H_0: $\beta_1 = 0$ against a one-sided or a two-sided alternative depending on the investigator's hypothesis of establishing a change or the direction of the change of the Y values. If the investigator wants to establish that Y changes with changing X values, the null and alternative hypotheses are

$$H_0: \beta_1 = 0, \qquad H_A: \beta_1 \neq 0$$

If the purpose of the study is to show that Y values increase with increasing X values, a one-sided test with the following null and alternative hypotheses is used:

$$H_0: \beta_1 \leq 0, \qquad H_A: \beta_1 > 0$$

On the other hand, if the investigator wants to know whether Y decreases as X increases, the null and alternative hypotheses are

$$H_0: \beta_1 \geq 0, \qquad H_A: \beta_1 < 0$$

The test statistic for testing the regression coefficient is the T-ratio given in the row labeled X. This T statistic has $n - 2$ df, where n is the number of pairs of observations. Some programs give the p value for this test statistic. However, the Minitab program does not give the p value and it should be calculated as described in the Appendix of chapter 4.

Instead of testing the significance of the population regression coefficient, an interval estimate of the population regression coefficient can also be obtained by using the equation in the Appendix, and draw the inferences.

For the problem at hand, it is logical to think that BP increases with age. To verify whether or not the data agree with this belief, one tests the null hypothesis H_0: $\beta_1 \leq 0$ against the alternative hypothesis H_A: $\beta_1 > 0$. For this purpose the test statistic from the computer printout is T-ratio $= 6.24$. The p value for this test (see Appendix, Chapter 4) is less than 0.005. This implies that if H_0 is true, there is less than a 0.005 chance of getting evidence against the null hypothesis that is as strong as the evidence at hand—or stronger, and such evidence is highly improbable. Thus H_0 is rejected and it is concluded that $\beta_1 > 0$; that is, the data suggest that BP increases with increasing age in men.

A 95% confidence interval for β_1 is 1.57 ± 0.57 (see Appendix). Thus, for each 1-year increase in age, the BP increases anywhere from $1.57 - 0.57 = 1.00$ to $1.57 + 0.57 = 2.14$.

R^2 Value and the Predicting Ability of the Equation

The R^2 value converted into percentage indicates the percentage of variability in the dependent variable explained by the regression equation. The R^2 value is also called the coefficient of determination. The higher the R^2 value, the better the predicting ability of the equation. The R^2 value in the printout is an estimate of the population parameter $\bar{R}^2$. If $\bar{R}^2 = 0$, the equation is not good and one predicts the dependent variable as $\bar{Y}$ irrespective of the X value. Testing $\bar{R}^2 = 0$ is equivalent to testing the regression coefficient (or regression coefficients, when there are several independent variables) equal to zero. In case of one independent variable, $\bar{R}^2 = 0$ can be tested by using the T-ratio from the row labeled X as discussed in the last subsection.

If the hypothesis $\bar{R}^2 = 0$ is rejected, it does not immediately imply that a good predicting equation was obtained. The *analysis of variance* table of the output will give more insight into the problem. The total variation present in the dependent variable Y will be partitioned into explainable and unexplainable components. The fitted $\hat{Y}$ values at the X values observed in the data explain a part of the variability in the dependent variable, and as these $\hat{Y}$ values are on the regression equation, this explained variation is ascribed to regression. How the points scatter around the fitted regression equation is unexplainable and purely random, and is called the residual variation. The variation due to different components is measured by the quantity called sum of squares (SS). The number of independent linear functions of the dependent variable which provide the necessary sum of squares is called the degrees of freedom (DF) and is a difficult concept for the students at the level of this book. The mean square (MS) is the ratio of SS to DF. Wetz (1964) suggests to use the observed F ratio of (regression MS)/(residual MS) in the analysis of variance portion of the computer output and to conclude the equation to have an excellent predicting capability, if it exceeds four times the critical value of the F ratio. The critical values of the F ratio will be discussed in the Appendix of this chapter.

In the present problem $R^2 = 81.2\%$ and thus 81.2% of the variability in BP is explained by the ages of the men. The observed F ratio is

$$F \text{ ratio} = \frac{1327.2}{34.1} = 38.92$$

The 0.05 level of significance critical F value in this context with 1 and 9 degrees of freedom is 5.12. Because 38.92 far exceeds four times 5.12, the predicting equation derived from the computer output based on current data, has excellent predicting ability for the BP of men given that their age is in the range of 38 to 60 years.

The R^2 value adjusted for df will be used later to compare two or more predicting equations using different numbers of independent variables.

Residual Plots and Outliers

The studentized residuals labelled TRESD given below the analysis of variance table in the computer output are the differences of the observed and predicted dependent variable values in terms of the standard deviations of such differences. If any studentized residual is below -3 or above 3, the observation associated with that residual is considered an outlier, in the sense that it does not belong to the particular body of data. For the problem under discussion, no studentized residual is below -3 or above 3, and hence there are no outliers in the data set of Table 5.3. If any outlier is detected, it will be dropped from the data, and a new equation is fitted for the remaining data.

The validity of the assumptions made to derive the regression equation can be examined through the residual plots. There are several types of residual plots, and the plot of studentized residual versus the X variable is exhibited in Figure 5.8. The plot does not show any undue

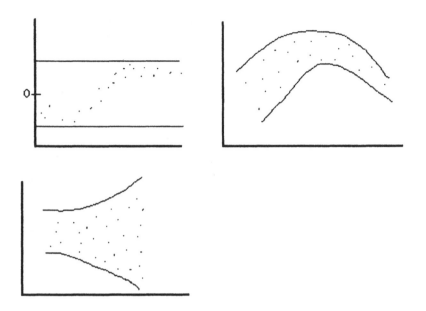

Figure 5.9 Bad residual plots.

pattern and hence the assumptions made in the course of this statistical analysis are reasonably justified.

If the points are scattered in a band with both positive and negative residuals occurring almost equally often, the assumptions made are reasonably valid. If it has any of the forms shown in Figure 5.9 or their variations, the assumptions are not tenable.

Confidence and Prediction Intervals

These are interval estimates of the dependent variable for a given value of the independent variable. To estimate the average of the dependent variable of all units with the given X value, a confidence interval is used. To estimate the dependent variable for any one selected unit with the given X value, a prediction interval is used. The prediction interval is wider than the confidence interval, because individual observations fluctuate more widely than do the means. The formulae are given in the Appendix and they use the S value given in the printout. The S value is called the standard error of the estimate and is the positive square root of residual MS.

To predict the average of BPs of all males aged 40, the confidence interval will be used. The required 95% confidence interval (see Appendix) is 119 ± 6.95. Thus the average BP of all males aged 40 will be anywhere between 119 − 6.95 = 112.05 and 119 + 6.95 = 125.95, unless the sample is one of the 5% cases where the estimate is wrong. If Bill is a 40-year-old man, his BP will not be predicted by a confidence interval but by a prediction interval. A 95% prediction interval (see Appendix) here is 119 ± 14.93. Thus Bill's BP is somewhere between 119 − 14.93 = 104.07 and 119 + 14.93 = 133.93 unless he is one of the 5% of people in whom the estimate does not work.

Example 5.4 (continued). The Minitab output fitting a linear regression equation of Y on X is given below:

THE REGRESSION EQUATION IS

$\hat{Y} = 3.1 + 38.6\,X$

COLUMN	COEFFICIENT	ST. DEV. OF COEF.	T−RATIO = COEF/S.D.
	3.07	19.59	0.16
X	38.643	5.433	7.11

S = 28.75
R−SQUARED = 91.0 PERCENT
R−SQUARED = 89.2 PERCENT, ADJUSTED FOR D.F.

ANALYSIS OF VARIANCE

DUE TO	DF	SS	MS=SS/DF
REGRESSION	1	41812	41812
RESIDUAL	5	4132	826
TOTAL	6	45944	

		TRESD
X	Y	
0	32	1.37481
1	47	0.21755
2	65	-0.58940
3	92	-1.01442
4	132	-0.98416
5	190	-0.25870
6	275	1.90437

The linear regression equation to predict Y for given X is $\hat{Y} = 3.1 +$ 38.6X. The R^2 value for this fit is 91%. Thus 91% of the variability in Y is explained by the linear regression equation. The observed F ratio is

$$F \text{ ratio} = \frac{41812}{86} = 486.19$$

and it far exceeds four times the 0.05 level critical F value, 6.61. Thus, the equation has an excellent predicting ability. None of the studentized residuals is less than -3 or greater than 3 and, hence, there are no outliers. The point estimate of the estimated number of bacteria at, say, 1.5 hours is $\hat{Y} = 3.1 + (38.6)(1.5) = 61$, and a 95% prediction interval is 61 ± 82 (see Appendix). Thus in any colony at 1.5 hours, the number of bacteria observed varies between $61 - 82 = -21$ (or more appropriately 0) to $61 + 82 = 143$. The residual plot given in Figure 5.10 shows some curvature.

On the other hand, one can transform Y to Z given by $Z = \ln Y$ and fit a linear regression equation to predict Z for a given value of X. The Minitab printout obtained here is a consequence of reading the Y and X data in columns C1 and C2 and using the following Minitab commands:

```
MTB > name c1 is 'y'
MTB > name c2 is 'x'
MTB > let c3 = loge (c1)
MTB > name c3 is 'z'
MTB > regress c3 (1) c1 (store studentized residuals in
        c4)
MTB > name c4 is 'stresd'
MTB > print c2 - c4
MTB > plot c4 vs c2
```

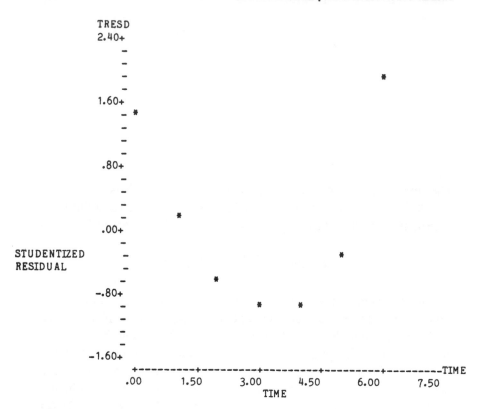

Figure 5.10   Residual plot.

THE REGRESSION EQUATION IS

$$\hat{Z} = 3.47 + 0.356\ X$$

COLUMN	COEFFICIENT	ST. DEV. OF COEF.	T−RATIO = COEF/S.D.
	3.47031	0.01037	334.54
X	0.355545	0.002877	123.58

S = 0.01522

R−SQUARED = 100.0 PERCENT
R−SQUARED = 100.0 PERCENT, ADJUSTED FOR D.F.

ANALYSIS OF VARIANCE

DUE TO	DF	SS	MS=SS/DF
REGRESSION	1	3.5396	3.5396
RESIDUAL	5	0.0012	0.0002
TOTAL	6	3.5407	

X	Z	STRESD
0	3.46574	−0.41090
1	3.85015	1.88766
2	4.17439	−0.50864
3	4.52179	−1.07576
4	4.88280	−0.70261
5	5.24702	−0.07910
6	5.61677	1.18317

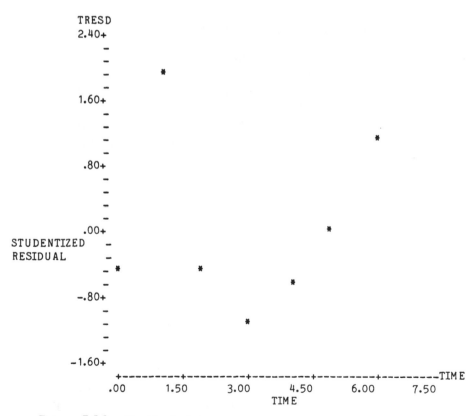

Figure 5.11   Residual plot.

The linear regression equation to predict Z for a given X is $\hat{Z} = 3.47 + 0.356X$. Noting that $\hat{Y} = e^{\hat{Z}}$, under this model $\hat{Y} = (32.14)(1.43)^X$. The R^2 value for the linear regression equation Z on X is 100.0%, and this equation explains practically all of the variation in Z. The equation also has an excellent predicting ability. All of the studentized residuals labeled STRESD in the output are between -3 and 3, and thus there are no outliers. The residual plot, Figure 5.11, has slightly removed the curvature that was present in Figure 5.10. The point estimate of the estimated number of bacteria at 1.5 hours is $\hat{Y} = (32.14)(1.43)^{1.5} = 54.96$. A 95% prediction interval on Z when $X = 1.5$ is 4.004 ± 0.043. Thus Z is between 3.961 and 4.047, and hence Y is between $e^{3.961} = 53$ to $e^{4.047} = 57$. The confidence and/or prediction intervals will first be calculated in the transformed scale and then changed back to the original scale.

POLYNOMIAL REGRESSION EQUATION

When the scatter plot Y versus X indicates a parabolic relation, i.e., a curve either opening upward or downward, a quadratic regression equation $\hat{Y} = b_0 + b_1X + b_2X^2$ will be fitted to predict Y by $\hat{Y}$, for a given X value.

In Minitab the data on the variables Y and X will be read in C1 and C2. Naming of the columns will be done by the following two commands:

```
MTB > name c1 is 'y'
MTB > name c2 is 'x'
```

Since X^2 is used in the model, there is a need to create a column C3 for X^2. This can be achieved by the command

```
MTB > let c3 = c2**2
```

The column C3 will now be labeled by the command

```
MTB > name c3 is 'xsq'
```

The necessary output for drawing inferences and diagnostic checks can be obtained with the help of the following three commands excluding the name command.

```
MTB > regress cl(2)c2, c3(store studentized residuals
      in c4)
MTB > name c4 is 'tresd'
MTB > print c2, c1, c4
MTB > plot c4 vs c2
```

Example 5.5 (continued). The Minitab output for the data of Table 5.5 using the above described Minitab commands is exhibited below using $\hat{Y}$ in place of Y.

THE REGRESSION EQUATION IS

$$\hat{Y} = 5.69 + 1.31 X - 0.0564 XSQ$$

COLUMN	COEFFICIENT	ST. DEV. OF COEF.	T-RATIO = COEF/S.D.
	5.685	3.187	1.78
X	1.3088	0.5578	2.35
XSq	-0.05636	0.02210	-2.55

S = 1.371

R-SQUARED = 58.9 PERCENT
R-SQUARED = 42.5 PERCENT, ADJUSTED FOR D.F.

ANALYSIS OF VARIANCE

DUE TO	DF	SS	MS=SS/DF
REGRESSION	2	13.472	6.736
RESIDUAL	5	9.403	1.881
TOTAL	7	22.875	

X	Y	TRESD
5	10	-1.00199
7	12	-0.07188
8	13	0.36578
10	15	1.51958
12	14	0.60964
13	12	-0.99794
15	11	-1.39604
20	10	1.94204

The fitted regression equation is $\hat{Y} = 5.69 + 1.31X - 0.0564 X^2$, and it can be used to predict the net profit/item, given the production in

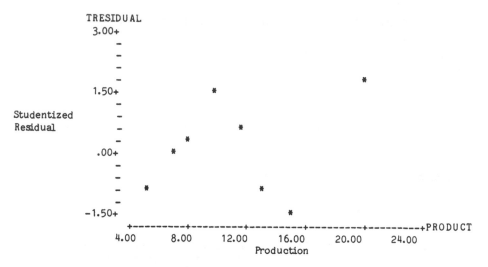

Figure 5.12   Residual plot.

thousands, X. If a company produces 9000 items, its profit/item can be determined as

$$\hat{Y} = 5.69 + (1.31)(9) - 0.0564(9^2) = \$12.91$$

The coefficients of X and X^2 terms in the estimated regression equation may be denoted by b_1 and b_2 respectively and are the estimates of the population parameters β_1 and β_2 in the model. If H_0: $\beta_1 = 0$, $\beta_2 = 0$ is tenable, the regression equation is not valid because the information on the X variable is not used to predict the Y value. On the other hand, if H_0 is rejected, then β_1 or β_2 is not zero, and the equation serves at least some purpose in predicting Y. This H_0 is tested by using the analysis of variance part of the output. The observed F ratio (regression MS)/(residual MS) from the output gives

$$F \text{ ratio} = \frac{6.736}{1.881} = 3.58$$

The p value for this F ratio as discussed in the Appendix is greater than 0.05. Since the observed p value is large, using $\alpha = 0.05$ we retain H_0 and conclude that a quadratic equation $\hat{Y} = b_0 + b_1 X + b_2 X^2$ does not give an adequate fit for the data. The scatter plot of Figure 5.12 indicates a third degree polynomial $\hat{Y} = b_0 + b_1 X + b_2 X^2 + b_3 X^3$ for consideration as a regression equation, because the curve is changing its direction twice.

The necessary Minitab commands to estimate and examine a third degree polynomial are the following ones in addition to the previous commands.

```
MTB > let c5 = c2 ** 3
MTB > name c5 is 'xcu'
MTB > regress cl(3) c2, c3, c5 (store studentized re-
        siduals in c6)
MTB > name c6 is 'stresd'
MTB > print c2, cl, c6
MTB > plot c6 vs c2
```

The following output will be obtained with the help of the above commands.

THE REGRESSION EQUATION IS

$$\hat{Y} = -10.6 + 6.18\,X - 0.486\,XSQ + 0.0114\,XCU$$

COLUMN	COEFFICIENT	ST. DEV. OF COEF.	T-RATIO = COEF/S.D.
	-10.622	6.078	-1.75
X	6.181	1.747	3.54
XSQ	-0.4861	0.1515	-3.21
XCU	0.011422	0.004009	2.85

S = 0.8808

R-SQUARED = 86.4 PERCENT
R-SQUARED = 76.3 PERCENT, ADJUSTED FOR DF

ANALYSIS OF VARIANCE

DUE TO	DF	SS	MS=SS/DF
REGRESSION	3	19.7715	6.5905
RESIDUAL	4	3.1035	0.7759
TOTAL	7	22.8750	

X	Y	STRESD	X	Y	STRESD
5	10	1.56538	12	14	0.93719
7	12	-1.02004	13	12	-0.91036
8	13	-0.79330	15	11	-0.45561
10	15	1.37935	20	10	1.01112

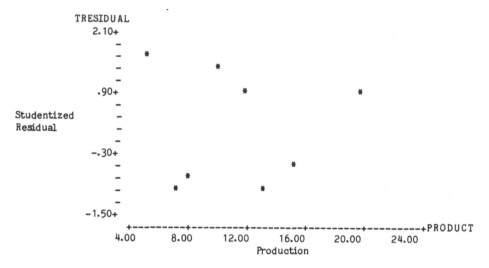

Figure 5.13   Residual plot.

The third degree polynomial equation predicting Y for a given X is $\hat{Y} =$ $-10.6 + 6.18X - 0.486X^2 + 0.0114X^3$. If a company produces 9000 items, its profit/item can be determined by this equation as

$$\hat{Y} = -10.6 + 6.18(9) - 0.486(9)^2 + 0.0114(9)^3 = \$13.96$$

Denoting the coefficients of X, X^2 and X^3 terms in the estimated regression equation by b_1, b_2, and b_3; they are the estimates of the population regression coefficients β_1, β_2, and β_3. The null hypothesis H_0: $\beta_1 = 0$, $\beta_2 = 0$, $\beta_3 = 0$ against the alternative hypothesis H_A: at least one β_i is not zero ($i = 1, 2, 3$), can be tested using the observed F ratio of (regression MS)/(residual MS) given by

$$F \text{ ratio} = \frac{6.5905}{0.7759} = 8.49$$

The p value for this F ratio is between 0.05 and 0.01. With $\alpha = 0.05$, H_0 is rejected and it is concluded that not all β_1, β_2, β_3 are zero. The equation predicts Y; but it is not an excellent predictor equation because the observed F ratio is less than four times the tabulated critical value. Since $R^2 = 86.4$ percent, it is concluded that 86.4% of the variability in Y is explained by the cubic regression.

Because the null hypothesis H_0: $\beta_1 = 0$, $\beta_2 = 0$, $\beta_3 = 0$, is rejected, one may want to see which of the regression coefficients are significant. In particular, one may like to test the significance of the β_3 coefficient in the model because it is an indicator of whether a cubic regression equation is needed or not. If the null hypothesis H_0: $\beta_3 = 0$ is rejected in favor of the alternative H_A: $\beta_3 \neq 0$, the third degree term in the model improves the prediction of Y; whereas, retention of H_0 makes the cubic term useless in the predicting equation. The test statistic T for testing H_0: $\beta_3 = 0$ is given in the row labeled XCU and is 2.85. The p value of this test statistic T for a two-sided test can be found as described in the Appendix of Chapter 4 and is between 0.02 and 0.05. Thus H_0: $\beta_3 = 0$ is rejected using $\alpha = 0.05$ in favor of H_A: $\beta_3 \neq 0$ and the third degree term is needed to improve the predicting potential of the equation. Each of the null hypothesis H_0: $\beta_1 = 0$ and H_0: $\beta_2 = 0$ can also be tested against two-sided alternatives and in this case they are also significant.

The studentized residuals labelled STRESD in the output can be examined for detecting any outliers in the data. Because none of the studentized residuals is less than -3 or greater than 3, there are no outliers in this data set.

The residual plot will be studied to determine whether or not the assumptions used to derive the equation are valid. Following the guidelines discussed earlier for the residual plot, one may agree (with a little hesitancy, because the plot is somewhat cone-shaped) that the assumptions are reasonably met.

One may predict Y through a prediction interval for a given X value if the prediction command exists in the version of Minitab used.

Working with a polynomial regression equation is a special case of working with a multiple regression equation as discussed in the next section.

MULTIPLE REGRESSION EQUATION

Let $X_1, X_2, \ldots, X_k$ be k independent variables and Y be the dependent variable. One is interested in predicting Y in a functional form of X_1, $X_2, \ldots, X_k$. Several functional forms can be considered. Some of the commonly used models are as follows:

1. $Y = \beta_0 + \beta_1 X_1 + \beta_2 X_2 + \cdots + \beta_k X_k + e$

2. $Y = \beta_0 + \beta_1 X_1 + \beta_2 X_2 + \cdots + \beta_k X_k$
    $\quad + \beta_{11} X_1^2 + \beta_{22} X_2^2 + \cdots + \beta_{kk} X_k^2 + e$

3. $Y = \beta_0 + \beta_1 X_1 + \beta_2 X_2 + \cdots + \beta_k X_k$
    $\quad + \beta_{12} X_1 X_2 + \beta_{13} X_1 X_3 + \cdots + \beta_{k-1,k} X_{k-1} X_k + e$

4. $Y = \beta_0 + \beta_1 X_1 + \beta_2 X_2 + \cdots + \beta_k X_k$
    $\quad + \beta_{11} X_1^2 + \beta_{22} X_2^2 + \cdots + \beta_{kk} X_k^2$
    $\quad + \beta_{12} X_1 X_2 + \beta_{13} X_1 X_3 + \cdots + \beta_{k-1,k} X_{k-1} X_k + e$

Depending on the nature of the independent variables and the form of their contributions on the dependent variable, the appropriate model is selected. To begin, the k independent variables selected in the study must not be highly correlated as pairs. If they are correlated, the resulting situation is known as *multicollinearity* and caution and different techniques should be used to cope with that situation. Such considerations are beyond the scope of this text. Each independent variable must be significantly correlated with the dependent variable.

Model 1 will be used when each independent variable is linearly related with the dependent variable when all other independent variables are held constant. It is called a *linear multiple regression equation*. Consider the example of predicting GPA(Y) of students in a semester based on their admission scores (X_1) and the average number of hours per week at work (X_2). Given the set of students with the same admission scores, it is reasonable to think a linear relationship for Y on X_2, that is, for each average hour per week work load, the student's GPA changes by a certain amount. Similarly, if the set of students are considered who work the same number of hours per week, then Y is linearly related to X_1 because for each point of admission score increase, one may expect an increase in GPA by a certain amount. With this logic in mind, one attempts to fit a model $Y = \beta_0 + \beta_1 X_1 + \beta_2 X_2 + e$ for this problem. The coefficients β_0, β_1, and β_2 will be estimated from the data as b_0, b_1 and b_2 and the sample regression equation $\hat{Y} = b_0 + b_1 X_1 + b_2 X_2$ can be obtained using Minitab package.

On all units, if the relationship of Y on each of the k variables is quadratic when all the remaining $k - 1$ variables are held constant, model 2 is appropriate.

Some variables interact in the sense that when both variables are used, the predicted Y value will not be the sum of the two predicted Y values using each of the individual independent variables, but will be higher or lower than that sum. Models 1 and 2 are considered as no interaction models while models 3 and 4 take interactions into account. When the investigator suspects interactions between the independent variables, models 3 and 4 are appropriate. Model 4 is considered as a complete *second degree multiple regression equation*.

Depending on the model used, in addition to the columns representing the independent variables, one develops columns of squares and products of the independent variables, as needed, in Minitab and regress the dependent variable on the independent variables. Some examples of the multiple regression equation will now be discussed.

Example 5.6. (Leaf Burn of Tobacco). The percentages of nitrogen (X_1), chlorine (X_2), potassium (X_3) and the log of leaf burn in seconds (Y) in 30 samples of tobacco from a farmers' fields are given in Table 5.6.

To fit a linear multiple regression equation of Y on X_1, X_2, X_3 in

Table 5.6 Percentage of Nitrogen (X_1), Chlorine (X_2), Potassium (X_3) and Log of Leafburn in Seconds (Y) in 30 Tobacco Samples

X_1	X_2	X_3	Y	X_1	X_2	X_3	Y
3.05	1.45	5.67	0.34	3.03	0.97	6.60	1.15
4.22	1.35	4.86	0.11	2.45	0.18	4.51	1.49
3.34	0.26	4.19	0.38	4.12	0.62	5.31	0.51
3.77	0.23	4.42	0.68	4.61	0.51	5.16	0.18
3.52	1.10	3.17	0.18	3.94	0.45	4.45	0.34
3.54	0.76	2.76	0.00	4.12	1.79	6.17	0.36
3.74	1.59	3.81	0.08	2.93	0.25	3.38	0.89
3.78	0.39	3.23	0.11	2.66	0.31	3.51	0.91
2.92	0.39	5.44	1.53	3.17	0.20	3.08	0.92
3.10	0.64	6.16	0.77	2.79	0.24	3.98	1.35
2.86	0.82	5.48	1.17	2.61	0.20	3.64	1.33
2.78	0.64	4.62	1.01	3.74	2.27	6.50	0.23
2.22	0.85	4.49	0.89	3.13	1.48	4.28	0.26
2.67	0.90	5.59	1.40	3.49	0.25	4.71	0.73
3.12	0.92	5.86	1.05	2.94	2.22	4.58	0.23

Source: Steel, R. G. H. and Torrie, J. H. (1960). *Principles and Procedures of Statistics*, McGraw-Hill, New York, p. 282 and their source O. J. Attoe.

Minitab, one reads the table data in columns C1, C2, C3, C4, and the following commands provide the necessary output where Y is used for Y:

```
MTB > name c1 is 'x1'
MTB > name c2 is 'x2'
MTB > name c3 is 'x3'
MTB > name c4 is 'y'
MTB > regress c4(3) c1-c3 (store studentized residuals
      in c5)
```

THE REGRESSION EQUATION IS

$$\hat{Y} = 1.81 - 0.531 \; X1 - 0.440 \; X2 + 0.209 \; X3$$

COLUMN	COEFFICIENT	ST. DEV. OF COEF.	T-RATIO = COEF/S.D.
	1.8110	0.2795	6.48
X1	-0.53146	0.06958	-7.64
X2	-0.43964	0.07304	-6.02
X3	0.20898	0.04064	5.14

S = 0.2135

R-SQUARED = 82.3 PERCENT
R-SQUARED = 80.2 PERCENT, ADJUSTED FOR DF

ANALYSIS OF VARIANCE

DUE TO	DF	SS	MS=SS/DF
REGRESSION	3	5.5047	1.8349
RESIDUAL	26	1.1848	0.0456
TOTAL	29	6.6895	

```
MTB > name c5 is 'tresd'
MTB > print c1-c5
```

X1	X2	X3	Y	TRESD
3.05	1.45	5.67	0.34	-1.95984
4.22	1.35	4.86	0.11	0.60267
3.34	0.26	4.19	0.38	-2.02147
3.77	0.23	4.42	0.68	0.24754

3.52	1.10	3.17	0.18	0.30893
3.54	0.76	2.76	0.00	-0.88465
3.74	1.59	3.81	0.08	0.82117
3.78	0.39	3.23	0.11	-0.98420
2.92	0.39	5.44	1.53	1.50896
3.10	0.64	6.16	0.77	-2.01126
2.86	0.82	5.48	1.17	0.45944
2.78	0.64	4.62	1.01	-0.03717
2.22	0.85	4.49	0.89	-1.55909
2.67	0.90	5.59	1.40	1.16514
3.12	0.92	5.86	1.05	0.37676
3.03	0.97	6.60	1.15	-0.01807
2.45	0.18	4.51	1.49	0.58982
4.12	0.62	5.31	0.51	0.26011
4.61	0.51	5.16	0.18	-0.19198
3.94	0.45	4.45	0.34	-0.54006
4.12	1.79	6.17	0.36	1.23010
2.93	0.25	3.38	0.89	0.19557
2.66	0.31	3.51	0.91	-0.41936
3.17	0.20	3.08	0.92	1.18490
2.79	0.24	3.98	1.35	1.44404
2.61	0.20	3.64	1.33	1.15953
3.74	2.27	6.50	0.23	0.25184
3.13	1.48	4.28	0.26	-0.65250
3.49	0.25	4.71	0.73	-0.49178
2.94	2.22	4.58	0.23	0.00180

The fitted regression is

$$\hat{Y} = 1.81 - 0.531X_1 - 0.440X_2 + 0.209X_3$$

This equation can be used to predict the log of leaf burn, when X_1, X_2, X_3 are given. If $X_1 = 3.0$, $X_2 = 1.5$, $X_3 = 4.0$, then $\hat{Y} = 1.81 - 0.531(3) - 0.440(1.5) + 0.209(4) = 0.393$.

The model used here is $Y = \beta_0 + \beta_1X_1 + \beta_2X_2 + \beta_3X_3 + e$. The coefficient β_1 is estimated by b_1, the coefficient of X_1, and is -0.531. This implies that for all samples that have the same percentage of chlorine and same percentage of potassium, if nitrogen increases by 1 percentage point, the log of leaf burn in seconds decreases by 0.531. The coefficient β_2 is estimated by b_2, the coefficient of X_2, and is -0.440. Thus, for samples that have the same percentage of nitrogen and same percentage of potassium, if chlorine increases by 1 percent-age point, the log of leaf burn in seconds decreases by 0.440. β_3 is estimated by b_3, the coefficient of X_3, and is 0.209. This means that for samples that have same percentage of nitrogen and the same per-centage of chlorine, if potassium increases by 1 percentage point, log of leaf burn in seconds increases by 0.209. Note that the change of Y is considered an increase or a decrease depending on the sign of the corresponding regression coefficient.

The most important independent variable in the predictor equation is not necessarily the one with the largest absolute value for the coeffi-cient, but it is the variable that has the largest absolute value for the T ratio. Here, X_1 is the most important variable in the predictor equation.

One can test the null hypothesis H_0: $\beta_1 = 0$, $\beta_2 = 0$, $\beta_3 = 0$ against the alternative H_A: at least one $\beta_i \neq 0$, by using the observed F ratio (regression MS)/(residual MS) from the Analysis of Variance table of the output, which is

$$F \text{ ratio} = \frac{1.8349}{0.0456} = 40.24$$

The p value of this F ratio is less than 0.01 and there is less than 1% chance of getting current evidence—or stronger evidence against H_0, when H_0 is true. Thus it will be concluded that at least one of β_1, β_2, or β_3 is not zero. It may be noted that retention of the H_0 implies that the fitted regression equation is an inadequate predictor and rejection of H_0 implies that the predictor equation serves some purpose. In this particular instance, the predictor equation is an excellent one because the observed F-ratio far exceeds four times the critical F ratio, that is, $40.24 > 4(2.98)$.

The R^2 value is 82.3%. Thus 82.3% of the variation present in the dependent variable is explained by the regression equation. Testing the hypothesis H_0: $\beta_1 = 0$, $\beta_2 = 0$, $\beta_3 = 0$ is equivalent to testing the hypothesis that the population $\bar{R}^2$ value is zero. Rejecting that the population $\bar{R}^2$ value is zero is not enough to get a good predictor equation. It should have a large R^2 value also.

Sometimes one needs to compare different multiple regression equations using different numbers of independent variables. Such com-parisons can be made based on the R^2 (adjusted) value, which adjusts for the degrees of freedom in its equation. With equations using different number of independent variables to predict a dependent variable, one way of choosing an appropriate equation is to choose the one with the largest R^2 (adjusted) value.

Outliers in the data can be detected and omitted in fitting the regression equation by examining the studentized residuals. If any studentized residual exceeds 3 or is below -3, the observation that provides such a studentized residual is an outlier for that particular data set. There are no outliers in the present data.

It was noted before that each population regression coefficient β_i measures the change in the dependent variable for a unit change in the corresponding independent variable X_i, when all other independent variables are held constant. These are known as partial regression coefficients. It becomes necessary to draw inferences on these parameters and one may wish to test formulated hypotheses about these coefficients and/or set confidence intervals on them. Significance of the regression coefficients can be tested by using the T ratio of the computer outputs. These T ratios are based on the residual degrees of freedom of the output, i.e., 26. The p values can be found as described in the Appendix of Chapter 4.

To test H_0: $\beta_1 = 0$, against the alternative H_A: $\beta_1 \neq 0$, the test statistic and the p value are

$$T \text{ ratio} = -7.64, \qquad p \text{ value} < .01$$

If $\beta_1 = 0$ is valid then there is less than 0.01 chance of getting evidence against the null hypothesis $\beta_1 = 0$ as strong as the evidence at hand—or stronger, and such evidence is highly improbable. Thus, $\beta_1 \neq 0$ will be concluded. The 95% confidence interval for β_1 (from the formula of the Appendix) is -0.53 ± 0.14.

To test H_0: $\beta_2 = 0$ against the alternative H_A: $\beta_2 \neq 0$, the test statistic and the p value are

$$T \text{ ratio} = -6.02, \qquad p \text{ value} < .01$$

Using a similar argument as in the last paragraph, it will be concluded that $\beta_2 \neq 0$. The 95% confidence interval on β_2 is -0.44 ± 0.15.

To test H_0: $\beta_3 = 0$ against the alternative H_A: $\beta_3 \neq 0$, the test statistic and the p-value are

$$T \text{ ratio} = 5.14, \qquad p \text{ value} < .01$$

From the evidence it will be concluded that $\beta_3 \neq 0$. The 95% confidence interval on β_3 is 0.21 ± 0.08.

Calculation of prediction intervals for predicting the dependent variable for given values of the independent variables can be performed if such a version of Minitab is available. The manual calculation of such intervals is beyond the scope of this book.

Example 5.7. (Gas Mileage). A car manufacturer may want to predict the gas mileage (Y) of cars given the engine size (X_1), horsepower (X_2), barrels (X_3), weight (X_4), and time (X_5). In Table 5.7, data is given on the five independent and the dependent variables for 20 car models. Enter the data of Table 5.7 in columns C1–C6 in Minitab. The following Minitab commands give the required output as given here where $\hat{Y}$ is used for Y. The Minitab commands and output are both given below:

```
MTB > name c1 is 'x1'
MTB > name c2 is 'x2'
MTB > name c3 is 'x3'
MTB > name c4 is 'x4'
MTB > name c5 is 'x5'
MTB > name c6 is 'y'
MTB > regress c6(5) c1 - c5 (store studentized residuals
      in c7)
```

THE REGRESSION EQUATION IS

$$\hat{Y} = 35.4 - 0.0002 \text{ X1} - 0.0323 \text{ X2} - 0.719 \text{ X3} - 0.00331 \text{ X4} + 0.126 \text{ X6}$$

COLUMN	COEFFICIENT	ST. DEV. OF COEF.	T-RATIO = COEF/S.D.
	35.44	14.77	2.40
X1	-0.00020	0.02169	-0.01
X2	-0.03235	0.03370	-0.96
X3	-0.7194	0.9887	-0.73
X4	-0.003308	0.002334	-1.42
X5	0.1262	0.7974	0.16

S = 2.891

R-SQUARED = 85.2 PERCENT

Table 5.7 Data on X_1, X_2, X_3, X_4, X_5 and Y

X_1	X_2	X_3	X_4	X_5	Y
160	110	4	2620	16.46	21
160	110	4	2875	17.02	21
108	93	1	2320	18.61	22.8
258	110	1	3215	19.44	21.4
360	175	2	3440	17.02	18.7
225	105	1	3460	20.22	18.1
360	245	4	3570	15.84	14.3
146.7	62	2	3190	20	24.4
140.8	95	2	3150	22.9	22.8
167.6	123	4	3440	18.3	19.2
167.6	123	4	3440	18.9	17.8
275.8	180	3	4070	17.4	16.4
275.8	180	3	3730	17.6	17.3
275.8	180	3	3780	18	15.2
472	205	4	5250	17.98	10.4
460	215	4	5424	17.82	10.4
440	230	4	5345	17.42	14.7
78.7	66	1	2200	19.47	32.4
75.7	52	2	1615	18.52	30.4
71.1	65	1	1835	19.9	33.9

Source: Part of the data from Freund, R. J. and Minton, P. D., *Regression Methods—A Tool for Data Analysis*, Dekker, New York, 1979, p. 156. Reproduced with the permission of the authors and publisher.

R–SQUARED = 80.0 PERCENT, ADJUSTED FOR D.F.

ANALYSIS OF VARIANCE

DUE TO	DF	SS	MS=SS/DF
REGRESSION	5	675.73	135.15
RESIDUAL	14	117.03	8.36
TOTAL	19	792.76	

```
MTB > name c7 is 'tresd'
MTB > print cl-c6
```

X1	X2	X3	X4	X5	Y	TRESD
160.0	110	4	2620	16.46	21.0	-0.58204
160.0	110	4	2875	17.02	21.0	-0.24430
108.0	93	1	2320	18.61	22.8	-1.46180
258.0	110	1	3215	19.44	21.4	-0.60948
360.0	175	2	3440	17.02	18.7	-0.14573
225.0	105	1	3460	20.22	18.1	-1.67634
360.0	245	4	3570	15.84	14.3	-0.24572
146.7	62	2	3190	20.00	24.4	0.18713
140.8	95	2	3150	22.90	22.8	-0.40551
167.6	123	4	3440	18.30	19.2	-0.10952
167.6	123	4	3440	18.90	17.8	-0.69316
275.8	180	3	4070	17.40	16.4	0.10925
275.8	180	3	3730	17.60	17.3	0.00525
275.8	180	3	3780	18.00	15.2	-0.73487
472.0	205	4	5250	17.98	10.4	-0.14903
460.0	215	4	5424	17.82	10.4	0.23505
440.0	230	4	5345	17.42	14.7	2.05330
78.7	66	1	2200	19.47	32.4	1.79548
75.7	52	2	1615	18.52	30.4	0.46642
71.1	65	1	1835	19.90	33.9	1.88550

The regression equation is

$$\hat{Y} = 35.4 - 0.0002X_1 - 0.0323X_2 - 0.719X_3 - 0.00331X_4 + 0.126X_5$$

A car with a 160 engine size, 110 horsepower, 2 barrels, weighing 2000 lb., with time 25.0 gives a gas mileage of

$$\hat{Y} = 35.4 - 0.0002(160) - 0.0323(110) - 0.719(2)$$
$$- 0.00331(2000) + 0.126(25) = 26.9$$

The weight of the car (X_4) is the most important predictor variable in estimating the gas mileage because the T ratio in the X_4 row has the largest absolute value. Denoting the regression coefficients in the model by β_1, β_2, β_3, β_4 and β_5, the null hypothesis H_0: $\beta_1 = 0$, $\beta_2 = 0$, $\beta_3 = 0$, $\beta_4 = 0$, $\beta_5 = 0$ is tested against the alternative hypothesis

H_A: at least one β_i is not zero, by using the F ratio (regression MS)/(residual MS), which is,

$$F \text{ ratio} = \frac{135.15}{8.36} = 16.17$$

The p value for this F ratio is less than 0.01 and H_0 is rejected. It is concluded that there is at least one β_i which is not zero. However, checking from the output, none of the individual regression coefficients based on the T ratios is significant. Such situations are not uncommon in statistical literature. No individual regression coefficient can be pinpointed as significant; yet, there is a significant explanation provided by the independent variables on the dependent variable. In fact, the equation is an excellent predictor because the observed F ratio exceeds four times the critical F ratio. All studentized residuals are between -3 to $+3$ and hence there are no outliers. Here 85.2% of the variability in gas mileages is explained by the fitted equation. This chapter is intended to give the reader some basic ideas of the regression methods. These methods are widely used in all areas of research. Interested readers may get a deeper understanding of these techniques by consulting standard texts on regression analysis like Draper and Smith (1981), Weisberg (1980), Neter and Wasserman (1974).

REFERENCES

Draper, N. and Smith, H. (1981). *Applied Regression Analysis,* 2nd ed. Wiley, New York.

Freund, R.J. and Minton, P.D. (1979). *Regression Methods—A Tool for Data Analysis.* Dekker, New York.

Neter, J. and Wasserman, W. (1974). *Applied Linear Statistic Models.* Irwin, Homewood.

Steel, R.G.D. and Torrie, J.H. (1960). *Principles and Procedures of Statistics.* McGraw-Hill, New York.

Weisberg, S. (1980). *Applied Linear Regression.* Wiley, New York.

Wetz, J.M. (1964). Criteria for Judging Adequacy of Estimation by an Approximating Response Function. Unpublished Ph.D. Thesis, Wisconsin University.

APPENDIX

Correlation Coefficient

Let (X_i, Y_i) for $i = 1, 2, \ldots , n$ be the n pairs of observations of the sample. The sample correlation coefficient between X and Y is

$$r = \frac{SS_{XY}}{\sqrt{SS_{XX}}\,\sqrt{SS_{YY}}} \qquad (5A.1)$$

where

$$SS_{XX} = \sum_{i=1}^{n} X_i^2 - \frac{\left(\sum_{i=1}^{n} X_i\right)^2}{n}$$

$$SS_{YY} = \sum_{i=1}^{n} Y_i^2 - \frac{\left(\sum_{i=1}^{n} Y_i\right)^2}{n} \qquad (5A.2)$$

$$SS_{XY} = \sum_{i=1}^{n} X_i Y_i - \frac{\left(\sum_{i=1}^{n} X_i\right)\left(\sum_{i=1}^{n} Y_i\right)}{n}$$

Example 5A.1. For the data of Table 5.1,

$$\sum_{i=1}^{n} X_i = 64 + 75 + 72 + 71 + 58 + 68 + 72 + 74 + 80 + 80$$
$$= 714$$

$$\sum_{i=1}^{n} Y_i = 72 + 65 + 69 + 78 + 70 + 88 + 73 + 80 + 76 + 73$$
$$= 744$$

$$\sum_{i=1}^{n} X_i^2 = 64^2 + 75^2 + 72^2 + 71^2 + 58^2 + 68^2 + 72^2 + 74^2$$
$$+ 80^2 + 80^2 = 51{,}394$$

$$\sum_{i=1}^{n} Y_i^2 = 72^2 + 65^2 + 69^2 + 78^2 + 70^2 + 88^2 + 73^2 + 80^2$$
$$+ 76^2 + 73^2 = 55{,}732$$

$$\sum_{i=1}^{n} X_i Y_i = (64)(72) + (75)(65) + (72)(69) + (71)(78) + (58)(70)$$
$$+ (68)(88) + (72)(73) + (74)(80) + (80)(76)$$
$$+ (80)(73) = 53,129$$

$$SS_{XY} = 53,129 - \frac{(714)(744)}{10} = 7.4$$

$$SS_{XX} = 51,394 - \frac{714^2}{10} = 414.4$$

$$SS_{YY} = 55,732 - \frac{744^2}{10} = 378.4$$

$$r = \frac{7.4}{\sqrt{414.4}\ \sqrt{378.4}} = 0.019$$

Example 5A.2. For the data of Table 5.3,

$$\sum_{i=1}^{n} X_i = 56 + 42 + 60 + 49 + 55 + 48 + 38 + 42 + 60 + 50$$
$$+ 50 = 550$$

$$\sum_{i=1}^{n} Y_i = 147 + 125 + 140 + 135 + 146 + 136 + 110 + 115$$
$$+ 148 + 140 + 140 = 1,482$$

$$\sum_{i=1}^{n} X_i^2 = 56^2 + 42^2 + 60^2 + 49^2 + 55^2 + 48^2 + 38^2 + 42^2$$
$$+ 60^2 + 50^2 + 50^2 = 28,038$$

$$\sum_{i=1}^{n} Y_i^2 = 147^2 + 125^2 + 140^2 + 135^2 + 146^2 + 136^2 + 110^2$$
$$+ 115^2 + 148^2 + 140^2 + 140^2 = 201,300$$

$$\sum_{i=1}^{n} X_i Y_i = (56)(147) + (42)(125) + (60)(140) + (49)(135)$$
$$+ (55)(146) + (48)(136) + (38)(110) + (42)(115)$$
$$+ (60)(148) + (50)(140) + (50)(140) = 74,945$$

$$SS_{XY} = 74,945 - \frac{(550)(1482)}{11} = 845$$

$$SS_{XX} = 28,038 - \frac{550^2}{11} = 538$$

$$SS_{YY} = 201,300 - \frac{1482^2}{11} = 1634.1818$$

$$r = \frac{845}{\sqrt{538} \sqrt{1634.1818}} = 0.901$$

Simple Linear Regression Equation

Consider a set of n pairs of observations (X_i, Y_i) for $i = 1, 2, \ldots, n$, and let the scatter plot indicate a straight-line fit for data. For the type of analysis discussed in this book, it is assumed that the Y values are normally distributed with the same variance for each value of the independent variable and that the means of the Y values at each X value form a straight line (see Figure 5.7). If $\hat{Y} = b_0 + b_1 X$ is the fitted straight line, b_0 and b_1 are found from data, by the method of least squares, which consists of minimizing

$$\sum_{i=1}^{n} (Y_i - \hat{Y}_i)^2 = \sum_{i=1}^{n} (Y_i - b_0 - b_1 X_i)^2$$

with respect to the unknown quantities b_0 and b_1.

By calculus methods, it can be found that such b_0 and b_1 are given by

$$b_1 = \frac{SS_{XY}}{SS_{XX}}, \qquad b_0 = \bar{Y} - b_1 \bar{X} \qquad\qquad (5A.3)$$

where SS_{XY}, SS_{XX} are given in (5A.2), $\bar{X} = \sum_{i=1}^{n} X_i/n$, and $\bar{Y} =$

$\sum_{i=1}^{n} Y_i/n$.

The total variability present in the dependent variable Y, SS_{YY}, is called the total sum of squares (SS) with $n - 1$ degrees of freedom. It will consist of a component resulting from the fitted regression line, known as regression SS, and given by

$$\text{Regression SS} = b_1 SS_{XY} = SS_{XY}^2/SS_{XX} \qquad\qquad (5A.4)$$

and a residual part, given by

$$\text{Residual SS} = \text{Total SS} - \text{Regression SS} \qquad (5A.5)$$

Each component of the sum of squares (SS) divided by its degrees of freedom (df) is called a mean square (MS). The degrees of freedom for regression in a simple linear regression equation is 1, because the regression equation has only one regression coefficient b_1. Since 1 degree of freedom is accounted by regression in a total of $n - 1$ degrees of freedom, the residual has $n - 2$ degrees of freedom. The observed F ratio is (Regression MS)/(Residual MS) and it has a distribution known as F distribution. The F distribution uses a pair of degrees of freedom, known as numerator degree of freedom v_1, and denominator degree of freedom v_2. v_1 is the degrees of freedom of the numerator component and v_2 is the degrees of freedom of the denominator component of the F ratio. In Table 3, at the end of the book, critical values of 5% and 1% of the F ratios are given. $F_{\alpha;v_1,v_2}$ will denote the critical value at α level of significance of a F ratio with v_1 numerator and v_2 denominator degrees of freedom. For example $F_{.05;4,12} = 3.26$, $F_{.01;1,9} = 10.56$. The p value for the calculated F ratio can also be obtained from the F tables given at the end of this book. If F ratio exceeds $F_{.01;v_1,v_2}$, the p value is less than 0.01; if it is between $F_{.05;v_1,v_2}$ and $F_{.01;v_1,v_2}$, the p value is between .05 and .01; and if it is less than $F_{.05};v_1, v_2$, the p value is more than .05. One way of testing the null hypothesis H_0: $\beta_1 = 0$ against the alternative hypothesis H_A: $\beta_1 \neq 0$ is to reject H_0 when the observed F ratio of the analysis of variance table exceeds the critical F ratio or equivalently when the p value $< \alpha$. An alternative procedure for this test is to use a T-ratio as discussed.

The $\sqrt{\text{Residual MS}}$ is denoted by s and is the estimated common standard deviation of the Y values from the regression line at each given X value. The standard deviation of the coefficient b_1 written s_{b_1}, is given by the formula

$$\text{SD of Coef. } (b_1) = s_{b_1} = \frac{s}{\sqrt{SS_{XX}}} \qquad (5A.6)$$

The T ratio for testing H_0: $\beta_1 = 0$ is

$$T \text{ ratio} = \frac{b_1}{s_{b_1}} \qquad (5A.7)$$

and the p value for this T ratio with residual df can easily be calculated for a two-sided or one-sided alternative (see Appendix of Chapter 4).

A confidence interval on β_1 with confidence coefficient $1 - \alpha$ is

$$b_1 \pm (t_{\alpha/2,\nu_2})(s_{b_1}) \qquad (5A.8)$$

where ν_2 is the residual degrees of freedom.

The R^2 value, also known as the coefficient of determination, is given by the formula

$$R^2 = \frac{\text{Regression SS}}{\text{Total SS}} \qquad (5A.9)$$

From (5A.9), it is clear that $0 \leq R^2 \leq 1$.

A $(1 - \alpha)$ 100% confidence interval on the estimated mean of Y values when $X = X_p$ is

$$\hat{Y}_p \pm (t_{\alpha/2,\nu_2})(s) \sqrt{\frac{1}{n} + \frac{(X_p - \bar{X})^2}{SS_{XX}}} \qquad (5A.10)$$

where

$$\hat{Y}_p = b_0 + b_1 X_p \qquad (5A.11)$$

A $(1 - \alpha)$ 100% prediction interval for the estimated individual Y values when $X = X_p$ is

$$\hat{Y}_p \pm (t_{\alpha/2,\nu_2})(s) \sqrt{1 + \frac{1}{n} + \frac{(X_p - \bar{X})^2}{SS_{XX}}} \qquad (5A.12)$$

Comparing (5A.10) and (5A.12), one notes that prediction intervals used to predict individual Y values are wider than the confidence intervals used to predict the mean of Y values. Furthermore, the least width of confidence and prediction intervals occur when $X_p = \bar{X}$. It may be recalled that $Y_p = \bar{Y}$ when $X_p = \bar{X}$.

Example 5A.2 (continued).   Consider the data of Table 5.3 again. Clearly,

$$\bar{X} = \frac{550}{11} = 50, \qquad \bar{Y} = \frac{1482}{11} = 134.73$$

From equation (5A.3),

$$b_1 = \frac{845}{538} = 1.5706, \qquad b_0 = 134.73 - (1.5706)(50) = 56.20$$

Thus, the fitted regression equation is

$$\hat{Y} = 56.2 + 1.57X$$

From equations (5A.4) and (5A.5)

$$\text{Regression SS} = \frac{845^2}{538} = 1327.18$$

$$\text{Residual SS} = 1634.18 - 1327.18 = 307.0$$

The analysis of variance (ANOVA) is then summarized in the following table. Clearly $s = \sqrt{34.1} = 5.84$.
From (5A.6),

$$s_{b_1} = \frac{5.84}{\sqrt{538}} = 0.2518$$

Table 5A.1   ANOVA

Due to	DF	SS	MS = SS/DF
Regression	1	1327.2	$\frac{1327.2}{1} - 1327.2$
Residual	9	307.0	$\frac{307.0}{9} = 34.1$
Total	10	1634.2	

The T ratio to test the hypothesis H_0: $\beta_1 = 0$, against one-sided or two-sided alternatives from equation (5A.7) is

$$T \text{ ratio} = \frac{1.5706}{0.2518} = 6.24$$

A 95% confidence interval on β_1 is $1.57 \pm (2.262)(0.2518) = 1.57 \pm 0.57$. The R^2 is given by (5A.9) and is

$$R^2 = \frac{1327.2}{1634.2} = 0.812$$

To get confidence and prediction intervals for Y when $X = 40$, one first calculates

$$\hat{Y}_p = 56.2 + (1.57)(40) = 119$$

A 95% confidence interval on the mean BP for all men aged 40, can be calculated from equation (5A.10), and is

$$119 \pm (2.262)(5.84)\left(\sqrt{\frac{1}{11} + \frac{(40 - 50)^2}{538}} \right) = 119 \pm 6.95$$

A 95% prediction interval for BP of any man aged 40, can be calculated from formula (5A.12), and is

$$119 \pm (2.262)(5.84)\left(\sqrt{1 + \frac{1}{11} + \frac{(40 - 50)^2}{538}} \right) = 119 \pm 14.93$$

These are the quantities given in a part of the computer output and used in the discussion of the main text.

Example 5A.3.   For the data of Table 5.4, clearly

$$\bar{X} = 3, \qquad \bar{Y} = 119,$$
$$SS_{XX} = 28, \qquad SS_{YY} = 45{,}944, \qquad SS_{XY} = 1{,}082$$

Hence

$$b_1 = \frac{1082}{28} = 38.643, \qquad b_0 = 119 - (38.643)(3) = 3.07$$

The regression equation is

$$\hat{Y} = 3.07 + 38.643X$$

$$\text{Regression SS} = \frac{1{,}082^2}{28} = 41{,}812$$

$$\text{Residual SS} = 45{,}944 - 41{,}812 = 4{,}132$$

The ANOVA can then be set as in Table 5A.2.
   Clearly $s = \sqrt{826.4} = 28.75$.

$$s_{b_1} = \frac{28.75}{\sqrt{28}} = 5.433$$

The T ratio to test H_0: $\beta_1 = 0$ against one-sided or two-sided
alternatives is

$$T \text{ ratio} = \frac{38.643}{5.433} = 7.11$$

A 95% confidence interval on β_1 is $38.643 \pm (2.571)(5.433) = 38.643 \pm 13.968$. R^2 is given by

$$R^2 = \frac{41{,}812}{45{,}944} = 0.91$$

Table 5A.2   ANOVA

Due to	DF	SS	MS = SS/DF
Regression	1	41,812	$\frac{41{,}812}{1} = 41{,}812$
Residual	5	4,132	$\frac{4{,}132}{5} = 826.4$
Total	6	45,944	

To get confidence and prediction interval for Y when $X = 1.5$, one first calculates

$$\hat{Y}_p = 3.07 + (38.643)(1.5) = 61$$

A 95% confidence interval on the mean Y when $X = 1.5$ is

$$61 \pm (2.571)(28.75)\left(\sqrt{\frac{1}{7} + \frac{(1.5 - 3)^2}{28}} \right) = 61 \pm 34$$

A 95% prediction interval on the individual Y when $X = 1.5$ is

$$61 \pm (2.571)(28.75)\left(\sqrt{1 + \frac{1}{7} + \frac{(1.5 - 3)^2}{28}} \right) = 61 \pm 82$$

Confidence Intervals on Regression Coefficients in a Multiple Regression Equation

Based on the computer output, the confidence interval on β_i with confidence coefficient $1 - \alpha$ is

$$b_i \pm (t_{\alpha/2, v_2}) s_{b_i} \qquad (5A.13)$$

where v_2 is the residual degrees of freedom in the analysis of variance table, b_i are given in column 2 and s_{b_i} are given in column 3. b_i and s_{b_i} appear in the columns headed by COEFFICIENT and ST. DEV. OF COEF. of the output.

Example 5A.4.   Consider the output of Example 5.6. A 95% confidence interval on the regression coefficient β_1 is

$$-0.53 \pm (t_{.025,26})(0.06958) = -0.53 \pm (2.056)(0.06958)$$
$$= -0.53 \pm 0.14$$

A 95% confidence interval on β_2 is

$$-0.44 \pm (2.056)(0.07304) = -0.44 \pm 0.15$$

A 95% confidence interval on β_3 is

$$0.21 \pm (2.056)(0.04064) = 0.21 \pm 0.08$$

EXERCISES

1. For 12 carrots, the length in cm (X) and weight in grams (Y) were studied and the following quantities were calculated:

$\bar{X} = 20,$ $\quad \bar{Y} = 30,$ $\quad SS_{YY} = 996,$
$SS_{XX} = 800,$ $\quad SS_{XY} = 760$

   (a) Calculate the correlation coefficient between X and Y.
   (b) Find the linear regression equation to predict Y given X.
   (c) Interpret the estimated regression coefficient and find the 95% confidence interval for the population regression coefficient.
   (d) Find an interval estimate with 0.95 probability for average weight, when length is 18 cm.

2. For 10 people, the height (in inches) X, and the chest measurement (in inches) Y were noted and the following quantities were calculated:

$\bar{X} = 69.2,$ $\quad \bar{Y} = 41,$ $\quad SS_{XX} = 25.6,$ $\quad\quad\quad SS_{YY} = 34,$
$SS_{XY} = 24$

   (a) Calculate the correlation coefficient between X and Y.
   (b) Find the linear regression equation to predict Y given X.
   (c) Test whether the population regression coefficient is zero.
   (d) Is the equation obtained, a good predictor equation?
   (e) Find a 95% interval estimate to predict the chest size of Bob who is 70 inches tall.

For the following problems, obtain a Minitab output and interpret the results as in the text. (Note that in some versions you may not get confidence interval and prediction interval to estimate the mean or individual Y values.)

3. The following table gives the population in 1970 (X) and the population in 1980 (Y) in selected cities of Nevada.

Place	Population in 1980 (Y)	Population in 1970 (X)
Boulder City	9,590	5,223
Carson City	32,022	15,468
East Las Vegas	6,449	6,501
Elko	8,758	7,621
Henderson	24,363	16,395
Las Vegas	164,674	125,787
Nellis A.F.B.	6,205	6,449
North Las Vegas	42,739	46,067
Paradise	84,818	24,477
Reno	100,756	72,863
Sparks	40,780	24,187
Sunrise Manor	44,155	9,684
Sun Valley	8,822	2,414
Winchester	19,728	13,981

Source: The World Almanac and Book of Facts, 1985, p. 272.

Regress Y on X.

4. The artificial data on percentage protein content (Y) and yield in pounds (X) of wheat from 10 agricultural plots are given in the following table.

Plot no.	Y	X			
1	10.7	22	6	9.8	18
2	10.8	21	7	10.1	17
3	10.8	19	8	10.4	17
4	10.2	19	9	10.3	17
5	10.3	18	10	10.9	16

Fit a regression equation $\hat{Y} = b_0 + b_1 X + b_2 X^2$.

5. The scores obtained in the first test (X_1), second test (X_2) and the final examination (Y) for 10 students in an elementary course are given in the following table.

Student #	X_1	X_2	Y
1	63	75	53
2	90	86	75
3	70	73	80
4	93	90	90
5	80	50	62
6	80	100	75
7	75	33	60
8	60	40	80
9	93	90	60
10	90	85	75

Fit a regression equation $\hat{Y} = b_0 + b_1X_1 + b_2X_2$.

Answers

(1a) 0.85; (1b) $\hat{Y} = 11 + 0.95X$; (1c) The weight of carrots on an average increase by 0.95 grams for each 1 cm. increase in length; 0.95 ± 0.41; (1d) 28.1 ± 3.55; (2a) .81; (2b) $\hat{Y} = -23.88 + 0.9375X$; (2c) T-ratio $= 3.96$, p-value $< .01$; (2d) observed F-ratio $= 15.65$, a good predictor, but not excellent; (2e) 41.75 ± 2.93.

6

Why Do Analysis of Variance?

An analysis of variance consists of partitioning the total variation in a variable into meaningful components and then drawing inferences about the respective components. This technique was used in the last chapter to partition the total variation in the dependent variable into regression and residual components, and inferences were drawn on the regression coefficients, and the adequacy of the fitted regression equation.

Whenever it is possible to isolate the causes of variability, an analysis of variance is performed and inferences are drawn. This is a widely used technique in statistically analyzing data. In this chapter this tool will be used to compare the equality of means of two or more populations. While it is possible to use a Minitab package of analysis of variance to compare the means of two or more populations, most textbooks discuss two-population problems separately, because it gives insight into the basics and uses analysis of variance to compare more than two-population means. In the main body of this chapter, this technique will be discussed through a computer package for two or more populations. In the chapter Appendix, the mathematical details of the two-population problem will be first considered and analysis of variance techniques will be discussed for more than two populations.

ONE-WAY ANALYSIS OF VARIANCE (FIXED EFFECTS)

There are two scenarios in which a one-way analysis of variance is used. When there are k distinct populations, inferences on all k population means can be made by drawing k independent samples from k populations and making the analysis of variance by splitting the total variability into between components and within components. Alternatively, if an experimenter is interested in testing the equality of effects of k test treatments, a set of homogeneous experimental units (EU) will be randomly divided into k groups, and each EU in the ith group will be given the ith test treatment for $i = 1,2, \ldots , k$. This design known as a Completely Randomized Design, was introduced in Chapter 2. The response measured from all EUs will be partitioned into treatment and error components. The treatment component explains the variability in the responses due to the applied treatments, and the error is the unexplained component of the variation. The error here is not a measurement error. Identical units treated alike may still indicate some differences in responses because of some unknown causes, and this variability is explained by the error component.

Usually the analysis performed on k samples is called the one-way analysis of variance and the analysis made on the responses of experimental units is called the analysis of a completely randomized design (CRD). The arithmetic used for these two analyses is the same and is appended to this chapter. Several examples of this analysis will be discussed now.

Example 6.1. (Television Watching Habits of Different Groups). Suppose a social scientist is interested in examining whether there is any difference between ethnic groups in the average number of hours of television watched per day. For this purpose, suppose four ethnic groups of people are considered: A, B, C, and D. Independent random samples of 10 people will be taken from each of the four groups and the data given in Table 6.1 are obtained.

The variation of the data for the forty people is due to the explainable difference between ethnic groups and the unexplainable difference

Table 6.1   Artificial Data for the Average
Number of Hours of Watching Television
per Day by Four Groups of People

	Groups			
	A	B	C	D
	3.5	4.1	5.0	2.8
	3.0	1.5	4.4	3.7
	4.5	3.4	4.4	1.6
	4.0	3.5	5.4	2.3
	4.4	3.5	3.1	2.8
	3.9	3.5	4.5	3.5
	4.2	3.0	4.8	3.9
	1.5	3.5	2.8	1.5
	2.3	3.3	3.4	3.0
	4.4	3.8	3.8	1.9
Means	3.57	3.31	4.16	2.70

within these groups. Thus, one may think of partitioning the total
variation into the components: 1. between groups, and 2. within groups.
If the variation among groups is considerably larger than the variation
within groups, then it is an indication of significant differences between
groups with respect to the average hours of television watching. The
partitioning of the variation is done through an analysis of variance.
Since there is only one explainable source of variation in this context, it
is called a one-way analysis of variance.

There are two different commands to do one-way analysis of variance on Minitab when the sample sizes are equal. The following command appears to be more convenient in such cases after reading the
four column data into C1-C4:

```
MTB > aovoneway for samples in c1-c4
```

The following output will then be obtained.

ANALYSIS OF VARIANCE

SOURCE	DF	SS	MS	F
FACTOR	3	10.997	3.666	4.93
ERROR	36	26.754	0.743	
TOTAL	39	37.751		

The factor component for this table is the between-the-groups component, and the error is the within the groups component. The null and alternative hypothesis tested from the analysis of variance output are

H_0: There are no differences among the four groups with respect to the average number of hours of watching television per day.

H_A: At least one pair among the four groups is different with respect to the average number of hours of watching television per day.

The test statistic for this purpose is the calculated $F = 4.93$ from the output. The p-value for this observed F based on the calculation as discussed in the Appendix of Chapter 5 is less than 0.01. There is less than 1% of a random chance of getting the present evidence—or stronger evidence, against H_0, when H_0 is true. Thus H_0 is rejected. This does not mean all the four ethnic groups are different with respect to watching television. It only means that at least one pair of groups differ in their television watching habits.

When H_0 is rejected indicating that at least two means are different, the natural question is to find which means are different. This problem is answered by multiple comparison methods. There are several procedures like LSD method, Tukey's T method, Duncan's multiple range test, Dunnett's method comparing treatments with control, and Scheffe's S method. The LSD Method will be discussed here while the other methods are beyond the scope of this book. An interested reader of the other methods may consult Federer (1967), Miller (1966), or Hochberg and Tamhane (1987). When the specific comparisons to be made between the groups are predetermined before data is collected and only a selected limited number of comparisons are planned, the least significant difference (LSD) method can be used. The LSD is a numerical value calculated by using the formula of the Appendix and is 0.76. If the investigator is primarily interested in comparing groups A and B, it can be concluded that there is no significant difference between those two groups in the average hours spent before a television, because the absolute difference of means for these groups, $|3.57 - 3.31| = 0.26$, is less than the LSD value 0.76. However, the groups C and D are significantly different in the average hours per day that they watch television, because the absolute difference between those groups is $|4.16 - 2.70| = 1.46$, and is greater than the LSD value.

The assumptions in performing the above analysis are:

1. The four samples are independently drawn from the respective populations. This assumption is met by the sampling procedure.
2. The data follow a normal distribution. This assumption can be tested by statistical techniques beyond the scope of this book. However, it is reasonable to assume normality for these data (as most people watch television a certain number of hours, e.g., two or three, and some people watch much longer, while some people watch much shorter times.)
3. The variances of the data within each group are same.

Assumptions such as these will be made in the context of analysis of the data discussed in the examples of this section.

Table 6.2 Artificial Data for Fasting Blood Sugar Levels of Volunteers Treated with Diabinese

	1 Tablet/day	2 Tablets/day	3 Tablets/day
	200	175	160
	195	180	160
	198	185	165
	200	175	165
	190	170	160
	195	180	165
	197	182	170
	195	180	165
	190	175	160
	195	180	170
Means	195.5	178.2	164.0

Example 6.2. (Treatment of Diabetes). Adult-onset diabetes can usually be treated, by diet, exercise, and oral medication. Suppose a physician is interested in studying the effect of three different dosages of the oral medication, Diabinese. A study can be planned using volunteers of same sex, age and diabetic condition. If 30 volunteers are available, they can be randomly divided into three groups. Each person in the first group will be given 1 tablet a day medication, each person in the second group will be given 2 tablets a day medication, and each person in the third group will be given 3 tablets a day medication. After a specific period of medication use, the fasting blood sugar levels can be measured as in Table 6.2.

Reading the data of Table 6.2 in C1-C3 and using the Minitab command

MTB > aovoneway for samples in c1-c3

the following output can be obtained:

ANALYSIS OF VARIANCE

SOURCE	DF	SS	MS	F
FACTOR	2	4977.3	2488.6	159.19
ERROR	27	422.1	15.6	
TOTAL	29	5399.4		

The factor component in the table is the treatment component in this example, and is the variation due to the tested treatments. The null and alternative hypotheses are

H_0:   The fasting blood sugar levels are the same for all dosages of the medication.

H_A: The fasting blood sugar levels are not all the same for the dosages of the medication.

The calculated test statistic is $F = 159.19$ with a p-value less than 0.01. The null hypothesis will be rejected and it will be concluded that there is a difference in fasting blood sugar levels for at least two dosages of medication. If the physician intends to examine the equality of the blood sugar levels with 1 tablet/day and 3 tablets/day he can do so by the LSD method. The LSD value (see Appendix for calculation) in this example is 3.6. The absolute difference in mean fasting blood sugar levels for volunteers receiving 1 tablet/day and those receiving 3 tablets/day is $|195.5 - 164.0| = 31.5$, and is larger than the LSD value, 3.6. Thus, the physician has evidence based on this data that 3 tablets/day yields a significantly different fasting blood sugar level than 1 tablet/day. It may be noted that the LSD method is a two-sided test. If the physician is interested in evidence that a higher dosage of medication reduces the blood sugar level, a one-sided test is more appropriate. The required test statistic (see Appendix) is

$$T = -17.8, \qquad p \text{ value} < .005$$

Under the hypothesis that the fasting blood sugar levels are identical with 1 tablet/day and 3 tablets/day, there is a probability of less than 0.5% of obtaining the observed—or stronger evidence against it. Such unlikely evidence is there; it will be concluded that 3 tablets/day is more effective in controlling fasting blood sugar than is 1 tablet/day in a similar patient group with identical physical and health profiles as had the volunteers used in the study.

Example 6.3. (Test Scores of Male and Female Students). In a class of 40 students of a first course in statistics there are 25 males. The test scores received by males and females are given in Table 6.3. The problem involved here is known as a two-sample (independent samples) T test method and can also be handled by the procedures of this section (for an alternative procedure, see Appendix). In the strict sense the samples are not independently drawn. However, it is fair to assume that the male and female students are reasonable representatives of their sexes and conclusions based on them are generally valid for the populations.

The interest here is to test the equality of population mean test scores of both males and females and we set the following:

H_0: Population Mean test scores of male and female students are the same.

H_A: Population Mean test scores of male and female students are not the same.

Since the sample sizes are different, the Minitab procedure discussed in the last two examples is not applicable here. The data should be

Table 6.3   Artificial Data of the Test
Scores of Male and Female Students

Males			Females	
80	70	57	73	79
76	61	63	55	74
85	65	82	84	76
90	75	71	52	82
89	89	61	90	83
84	57		61	
87	74		77	
60	61		72	
68	85		50	
89	78		63	

entered in Minitab in two columns C1 and C2. Data will be entered in C1, and the sample number in C2. Data of 25 males and 15 females will be entered in C1. The first 25 entries in C2 will be 1 and the last 15 entries in C2 will be 2. The Minitab command, "ONEWAY ON DATA IN C1, GROUPS IN C2", produces the necessary output. The commands typed by the user and the output obtained are given below:

```
MTB> read c₁, c₂
DATA> 80 1
DATA> 76 1
DATA> 85 1
DATA> 90 1
DATA> 89 1
DATA> 84 1
DATA> 87 1
DATA> 60 1
DATA> 68 1
DATA> 89 1
DATA> 70 1
DATA> 61 1
DATA> 65 1
```

```
DATA> 75 1
DATA> 89 1
DATA> 57 1
DATA> 74 1
DATA> 61 1
DATA> 85 1
DATA> 78 1
DATA> 57 1
DATA> 63 1
DATA> 82 1
DATA> 71 1
DATA> 61 1
DATA> 73 2
DATA> 55 2
DATA> 84 2
DATA> 52 2
DATA> 90 2
DATA> 61 2
DATA> 77 2
DATA> 72 2
DATA> 50 2
DATA> 63 2
DATA> 79 2
DATA> 74 2
DATA> 76 2
DATA> 82 2
DATA> 83 2
```

        40 ROWS READ

```
MTB> name c1 is 'score'
MTB> name c2 is 'sex'
MTB> oneway on data in c1, groups in c2
```

ANALYSIS OF VARIANCE ON SCORE

SOURCE	DF	SS	MS	F
SEX	1	78	78	0.56
ERROR	38	5259	138	
TOTAL	39	5336		

Sex in the output is between the sexes component, and error is within the sexes component. Whenever the calculated F value is less than 1, there is no need to look for the p-value and H_0 is retained automatically. Thus, based on these data, there is no evidence that male and female students get different mean test scores in a first statistics course. The confidence interval for the difference of population mean score of males (μ_1) and population mean score of females (μ_2) can be obtained through the following Minitab command:

```
MTB> twot [95 percent confidence] data in cl, groups in
     c2; pooled
95 PCT   SCORE   FOR   MU1 - MU2: (-4.9, 10.7)
```

The 95% confidence interval $\mu_1 - \mu_2$ is thus $-4.9 \leq \mu_1 - \mu_2 \leq 10.7$.

Example 6.4. (Pain Relievers). Suppose an investigator is interested in studying the effectiveness of common pain relievers for minor arthritic conditions. A group of volunteers with identical physical and health profiles may be selected for the study. Consider the pain relievers Bufferin, Excedrin, and Extra-Strength Tylenol for this study. If 30 volunteers are available for the experiment, they can be randomly divided into three groups each of 10 volunteers. Each volunteer in the first group will be assigned Bufferin, each volunteer of the second batch will be assigned Excedrin, and each volunteer of the third batch will be assigned Extra-Strength Tylenol. The time for relief (in minutes) after the medication is taken will be the response variable and artificial data in such an experiment is given in Table 6.4. Following the procedure discussed in Examples 6.1 and 6.2, the required Minitab output is the following:

```
ANALYSIS OF VARIANCE
```

SOURCE	DF	SS	MS	F
FACTOR	2	25.07	12.54	11.51
ERROR	27	29.40	1.09	
TOTAL	29	54.47		

The null and alternative hypotheses tested from the above computer output are

H_0:   The three pain relievers are equally effective with respect to the
          average time for relief.

H_A: The three pain relievers are not all equally effective with respect
          to the average time for relief.

The calculated F-value for this test from the output, is 11.51 with a
p-value less than 0.05. Since the p value is small, it will be concluded
that at least a pair of the three pain relievers are different with respect
to the average time the medication starts working. If the experimenter
is interested in comparing the effects of Bufferin and Extra-Strength
Tylenol, the required LSD value is 0.96. The absolute value of the
difference between the mean responses of these two treatments is
$|30.58 - 30.80| = .22$, and because this difference is smaller than the
LSD value, it will be concluded that Bufferin and Extra-Strength Tylenol provide relief at almost the same time the medication is taken by
the patient.

    In Examples 6.2 and 6.4 using an experimental design setting, the
experimenter drew inferences about the tested treatments only. However, in some experiments the interest lies in the entire population of

Table 6.4   Artificial Data on Time Required (in min)
for Pain Relief

	Bufferin	Excedrin	Extra-Strength Tylenol
	30.5	29.3	31.2
	30.1	27.4	31.0
	31.5	30.4	30.1
	31.0	27.1	30.5
	31.4	31.0	31.4
	30.9	28.1	29.9
	31.2	29.7	30.7
	28.5	29.2	30.6
	29.3	27.0	31.8
	31.4	28.4	30.8
Means	30.58	28.76	30.80

treatments and the experimenter will conduct an experiment using a random sample of treatments. In an animal-feeding experiment, if the investigator is interested in studying the equality of effects of all the available feeds in the market, it is not feasible to organize an experiment using all available treatments. Instead, a random sample of feeds will be selected and tested with the hope of generalizing experimental results that are applicable to the population of treatments.

When the interest of the investigator lies only in the tested treatments, the resulting analysis is based on a fixed effects model as discussed in the Appendix. The null and alternative hypotheses tested are based on the effects of tested treatments only. When the tested treatments are a random sample from a pool of treatments, the random effects model as discussed in the Appendix and the next section will be used to perform the analysis, and inferences will be drawn about the variance of the population treatment effects. The effects of tested treatments are inconsequential here.

ONE-WAY ANALYSIS OF VARIANCE (RANDOM EFFECTS)

For the purpose of this analysis, the treatments to be used in the experiment are randomly selected from a population of treatments and σ_T^2 is the variance of the population treatment effects. Usually one likes to estimate σ_T^2 rather than testing hypotheses about σ_T^2. Let σ_E^2 be the variance of the error components. The analysis of variance table can be obtained by using Minitab package in the last section. The mean square for the factor will not be the same if the experiment is repeated. The average value of this mean square can be mathematically calculated assuming that similar experiments are repeated many times and that the average is called the expected value of the factor mean square. A similar interpretation can be given to the expected value of the error mean square. The expected values of the mean squares will be written E(MS). By using (6A.19) and (6A.21) of the Appendix, the expected values of mean squares (E(MS)) will be calculated and will be entered in the analysis of variance table itself. By equating the E(MS)s to the respective MSs, σ_E^2 and σ_T^2 will be solved and these are the estimates $\hat{\sigma}_E^2$ and $\hat{\sigma}_T^2$ of the respective parameters. Examples will now be given of this analysis.

Example 6.5. (Chicken-Feeding Experiment). Four feeds were randomly selected from a population of chicken feeds and each was fed to five baby chicks. The response variable is the gain in weight (in pounds) after a specific period. The analysis of variance of artificial data of the variable is given below:

ANALYSIS OF VARIANCE

SOURCE	DF	SS	MS	E(MS)
FACTOR	3	65.59	21.86	$\sigma_E^2 + 5\sigma_T^2$
ERROR	16	28.90	1.81	σ_E^2
TOTAL	19	94.49		

From the two equations

$$\hat{\sigma}_E^2 = 1.81, \qquad \hat{\sigma}_E^2 + 5\hat{\sigma}_T^2 = 21.86$$

σ_T^2 will be solved as $\hat{\sigma}_T^2 = 4.01$. Thus, the variance of effects for gain in chicks' weights for population feeds is 4.01.

Example 6.6. (Breeding Experiments). Plant breeders develop a large number of new lines of a crop. It is practically impossible to test all of the new lines in one experiment. Thus, a random sample of lines usually will be experimented. Consider an experiment in which nine randomly selected lines are tested in a completely randomized design using four replications for each treatment. The response is the yield (bushels/acre). The following analysis of variance table is the computer output based on some artificial data.

ANALYSIS OF VARIANCE

SOURCE	DF	SS	MS	E(MS)
FACTOR	8	5927	741	$\sigma_E^2 + 9\sigma_T^2$
ERROR	27	4125	153	σ_E^2
TOTAL	35	10052		

From the two equations

$$\hat{\sigma}_E^2 = 153, \quad \text{and} \quad \hat{\sigma}_E^2 + 9\hat{\sigma}_T^2 = 741$$

$\hat{\sigma}_T^2$ will be solved as $\hat{\sigma}_T^2 = 65.3$. Thus, the variance of effects of yields of all the population lines of that crop is 65.3.

If the treatment effects are assumed to have a normal distribution, $6\sigma_T$ measures the range of the treatment effects. It may be noted that $\hat{\sigma}_T^2$ may turn out to be negative in this method of estimation, and in that case it is reasonable to assume it zero. It may be noted that there are other available methods providing positive estimates for σ_T^2.

RANDOMIZED BLOCK DESIGN

In an experimental setting, if all EUs are not homogeneous, they will be divided into homogeneous groups called blocks. The EUs in each block are assumed to be homogeneous. If each treatment appears once in each block, the resulting design is called a Randomized Block Design. It is important to randomly assign the treatments to EUs in each block using separate randomizations from block to block. The following examples illustrate the analysis of these designs.

Example 6.7. (Pain Relievers). Consider the experiment in Example 6.4 in which the investigator wants to test the effectiveness of the three pain relievers Bufferin, Excedrin, and Extra-Strength Tylenol. Suppose it is not possible to get enough volunteers with identical physical and health profiles to run a completely randomized design experiment. Instead, let there be three volunteers in each of the following eight groups: males 50–55 years age, males 55–60 years age, males 60–65 years age, males 65–70 years age, females 50–55 years age, females 55–60 years age, females 60–65 years age, and females 65–70 years age. These groups can be identified as eight blocks with three people in each block. In each of these groups the three pain relievers will be randomly assigned to the three volunteers in that group. Artificial data for the time (in minutes) that the relief begins after the medication is taken are given in Table 6.5. In this setting the variability of the data is due to three components: treatments, groups of volunteers and unexplained random error. The partitioning of total variation can be achieved through Minitab. The input is made in three columns. In column 1, the data are entered. In columns 2 and 3 the treatment and block numbers are entered for the corresponding response in column 1. The Minitab command TWOWAY ON DATA IN C1, SUBSCRIPTS IN C2, C3 provides the necessary output. The commands typed by the user and the output are as follows.

Table 6.5   Artificial Data for Time Required (in min) for Pain Relief When the Volunteers Are Grouped

Group	Bufferin	Excedrin	Extra-Strength Tylenol
Males 50–55	29.5	27.3	28.7
Males 55–60	29.6	29.0	30.4
Males 60–65	32.0	29.0	30.2
Males 65–70	32.0	28.6	33.0
Females 50–55	30.4	27.0	30.2
Females 55–60	30.4	28.5	27.3
Females 60–65	31.7	30.9	30.3
Females 65–70	29.5	28.2	30.7
Means	30.64	28.68	30.1

```
MTB> read cl-c3
MTB> 29.5    1    1
MTB> 27.3    2    1
MTB> 28.7    3    1
MTB> 29.6    1    2
MTB> 29.9    2    2
MTB> 30.4    3    2
MTB> 32.0    1    3
MTB> 29.0    2    3
MTB> 30.2    3    3
MTB> 32.0    1    4
MTB> 28.6    2    4
MTB> 33.0    3    4
MTB> 30.4    1    5
MTB> 27.0    2    5
MTB> 30.2    3    5
MTB> 30.4    1    6
MTB> 28.5    2    6
MTB> 27.3    3    6
MTB> 31.7    1    7
MTB> 30.9    2    7
```

```
MTB> 30.3    3    7
MTB> 29.5    1    8
MTB> 28.2    2    8
MTB> 30.7    3    8
```

      24 ROWS READ
```
MTB> name c1 is 'time'
MTB> name c2 is 'treat'
MTB> name c3 is 'block'
MTB> twoway on data in c1, subscripts in c2, c3
```

ANALYSIS OF VARIANCE ON TIME

SOURCE	DF	SS	MS
TREAT	2	16.46	8.23
BLOCK	7	21.02	3.00
ERROR	14	17.83	1.27
TOTAL	23	55.31	

While it is possible to test for the equality of block effects, the most commonly tested null and alternative hypotheses are

H_0: Treatments are equally effective with respect to the time of relief.

H_A: Treatments are not all equally effective with respect to the time of relief.

The test statistic for this purpose is the F value of (treat MS)/(Error MS) and is $8.23/1.27 = 6.48$. The p value for this statistic can be found as discussed in the Appendix of Chapter 5 and is between 0.05 and 0.01. Since the p value is less than $\alpha = 0.05$, H_0 is rejected at 0.05 level and it is concluded that the three medications differ in the time of the relief. If the experimenter wants to compare Bufferin and Extra-Strength Tylenol, the LSD value (see Appendix) is 1.21. The absolute mean difference for these medicines is $|30.64 - 30.1| = 0.54$ and this being smaller than the LSD value, it will be concluded that both these medicines have identical onset timing for the pain relief.

The analysis discussed in the above example and other examples in this section is based on the following assumptions:

1. The response variable is normally distributed.

2.   The variances of the response variable is same within each treatment and within each block.
3.   All observations are independently distributed.
4.   There is no treatment block interaction in the sense that the difference in the effects of medicines is uniformly same for all patient groups.

Even when condition 3 is not met, under some correlation structures of the response variable within each block, the analysis described in Example 6.7 is still valid.

Example 6.8.   (Paired Data). The protein content of 5 wheat varieties as determined by a standard and a newly developed rapid method gave the data in Table 6.6. The randomized block design analysis using only two treatments is also known as a paired T test. The analysis can be done using Minitab package as described in this section. It can also be performed by using the approach outlined in the Appendix. By using the Minitab commands as discussed in the last example, the following output can be obtained.

ANALYSIS OF VARIANCE ON PROTEIN

SOURCE	DF	SS	MS
METHOD	1	0.121	0.121
VARIET	4	11.390	2.847
ERROR	4	3.454	0.863
TOTAL	9	14.965	

Table 6.6   Artificial Data for Protein Content (g/100 g) in Five Wheat Varieties

Variety	Standard method	Rapid method
A	12.6	11.1
B	13.3	12.0
C	12.9	14.6
D	13.2	12.8
E	14.8	15.2
Means	13.36	13.14

The calculated F value for testing

H_0: Determinations of protein content based on standard and rapid methods are the same.

H_A: Determinations of protein content based on standard and rapid methods are not the same.

is 0.121/0.863 and is less than 1. Thus, H_0 is retained and it is concluded that the average determinations are same by both methods.

Example 6.9. (Gas Mileage). An investigator interested in studying the gas consumption at different speeds (mph) may conduct a randomized block design experiment. Suppose the interest lies in comparing gas mileages on a highway stretch at 45 mph, 50 mph, and 55 mph. Since different cars of the same model with similar equipment may differ in gas mileage delivered, each of the speeds should be tested on each experimental car. Let 10 cars be used for the study. The cars, however, need not all be identical. If different drivers are used, they should be switched only among the cars, but not on the same car. An artificial data from such an experiment is presented in Table 6.7.

Table 6.7 Artificial Data for Gas Mileage (miles/gallon) on 10 Cars at Different Speeds

Car	Speed		
	45 mph	50 mph	55 mph
1	25.5	23.5	21.2
2	25.1	22.6	21.8
3	26.5	23.0	23.7
4	26.0	23.0	22.6
5	26.4	25.4	21.6
6	25.9	22.2	23.2
7	26.2	23.9	23.2
8	23.5	22.7	25.3
9	24.3	25.0	23.0
10	26.4	24.3	21.9
Means	25.58	23.56	22.75

By proceeding as outlined earlier on Minitab, the following computer printout can be obtained.

```
ANALYSIS OF VARIANCE ON GASMIL

SOURCE      DF    SS        MS
SPEED       2     42.48     21.24
CAR         9      5.27      0.59
ERROR       18    27.30      1.52
TOTAL       29    75.05
```

The null and alternative hypotheses of interest in this problem are

H_0: The gas mileage is the same at three speeds.
H_A: The gas mileage for at least a pair of speeds is not same.

The test statistic is the F value $21.24/1.52 = 13.97$ with p value less than 0.01. Since the p value is small, H_0 is rejected and it is concluded that the gas mileage is not the same at all the three speeds. If the experimenter is interested to compare the gas mileages at 45 mph and 55 mph, the required LSD is 1.16. The absolute value of the difference between mean gas mileages at 45 mph and 55 mph is $|25.58 - 22.75| = 2.83$, and is larger than the LSD value. It will thus be concluded that the gas mileages delivered are different at 45 mph and 55 mph.

Analogous to the one-way analysis with random effects, here one can also select the blocks and treatments randomly from appropriate populations. Then one needs to know the expected value of mean squares to estimate the variance of treatment effects in the population.

Sometimes, treatments may be randomly selected from a pool of treatments while blocks are fixed, or one may select the blocks randomly and use a fixed set of treatments in the experiment. The resulting analysis is called a mixed effects analysis.

Random and mixed effects analysis will not be considered in this book. The interested reader is referred to Snedecor and Cochran (1980).

REFERENCES

Federer, W.T. (1967). *Experimental Design: Theory and Application.* Oxford & IBH, New Delhi.

Hochberg, Y. and Tamhane, A.C. (1987). *Multiple Comparisons.* Wiley, New York.

Miller, R.G., Jr. (1966). *Simultaneous Statistical Inference.* McGraw-Hill, New York.

Snedecor, G.W. and Cochran, W.G. (1980). *Statistical Methods.* 7th Ed. Iowa State University Press, Ames.

APPENDIX

Inferences on Two Population Means Based on Two Independent Samples

Suppose there are two populations. Let μ_1 and σ_1^2 be the mean and variance of the first population and let μ_2 and σ_2^2 be the mean and variance of the second population, where all the four parameters are unknown. The interest now is to test certain hypotheses about μ_1 and μ_2, and set confidence intervals on the difference $\mu_1 - \mu_2$. The problem is easily tractable if the populations are normally distributed and $\sigma_1^2 = \sigma_2^2$; only this case will be considered here. Let two independent random samples of sizes n_1 and n_2 be drawn from these two populations. Further let $\bar{X}_1$ and s_1^2 be the mean and variance of the first sample, and $\bar{X}_2$ and s_2^2 be the mean and variance of the second sample. Since each of s_1^2 and s_2^2 estimate the common unknown variance of the two populations, they can be pooled together to give a pooled variance, s_p^2, as indicated below:

$$s_p^2 = \frac{(n_1 - 1)s_1^2 + (n_2 - 1)s_2^2}{n_1 + n_2 - 2} \tag{6A.1}$$

The necessary test statistic to test H_0: $\mu_1 - \mu_2 = c$ against a one-sided or two-sided alternative is the T-statistic

$$T = \frac{\bar{X}_1 - \bar{X}_2 - c}{s_p \sqrt{\dfrac{1}{n_1} + \dfrac{1}{n_2}}} \tag{6A.2}$$

which has a student's T distribution with $n_1 + n_2 - 2$ degrees of freedom. Usually $c = 0$ for this test. A $(1 - \alpha)$ 100% confidence interval on $\mu_1 - \mu_2$ is

$$\bar{X}_1 - \bar{X}_2 \pm (t_{\alpha/2, n_1 + n_2 - 2}) \left(s_p \sqrt{\frac{1}{n_1} + \frac{1}{n_2}} \right) \qquad (6A.3)$$

Example 6A.1. Consider the problem discussed in Example 6.4. Let the male and female students be respectively designated groups 1 and 2. From the data given in Table 6.3, one obtains

$$n_1 = 25, \qquad \bar{X}_1 = 74.28, \qquad s_1^2 = 129.37667,$$
$$n_2 = 15, \qquad \bar{X}_2 = 71.4, \qquad s_2^2 = 153.82853$$

From formula (6A.1),

$$s_p^2 = \frac{(24)(129.37667) + (14)(153.82853)}{38} = 138.3853$$

Suppose μ_1 and μ_2 are the population mean scores for the two groups. The test statistic for testing $H_0: \mu_1 = \mu_2$ against $H_A: \mu_1 \neq \mu_2$, from formula (6A.2) is

$$T = \frac{74.28 - 71.4}{\sqrt{138.3853} \sqrt{\frac{1}{25} + \frac{1}{15}}} = 0.75$$

The degrees of freedom for this test statistic is 38. Since the T distribution with 38 degrees of freedom approximates the standard normal distribution,

$$p\text{-value} = 2P(Z > 0.75) = 0.4532$$

This p value is large and there is no evidence against H_0 from this data set. A 95% confidence interval can be calculated from (6A.3) and is

$$(74.28 - 71.4) \pm (1.96)\sqrt{138.38526} \sqrt{\frac{1}{25} + \frac{1}{15}} = 2.88 \pm 7.53$$

Thus $-4.65 \leq \mu_1 - \mu_2 \leq 10.41$. This interval is slightly different from the computer output given in Example 6.3 because, in the output, the critical value of an exact T distribution was used, while it is approximated here with the critical value of the standard normal variable. The absolute value of the test statistic given in (6A.2) with $c = 0$ is the square root of the F statistic of the analysis of variance discussed in Example 6.3.

Example 6A.2. Suppose a soil scientist is interested in studying the diffusion rates of carbon dioxide through fine soil and coarse soil. He can take independent sample observations of size 10 from each soil type. Suppose the following are the summary statistics:

$$n_1 = 10, \quad \bar{X}_1 = 24.4, \quad s_1^2 = 9.3778,$$
$$n_2 = 10, \quad \bar{X}_2 = 27.2, \quad s_2^2 = 17.0667$$

If μ_1 and μ_2 are the population means, the hypotheses of interest are

$$H_0: \quad \mu_1 = \mu_2, \quad H_A: \quad \mu_1 \neq \mu_2$$

The pooled variance for this data is the simple average of s_1^2 and s_2^2 as $n_1 = n_2$ and it is 13.22225. The test statistic is

$$T = \frac{24.4 - 27.2}{\sqrt{13.22225} \sqrt{\dfrac{1}{10} + \dfrac{1}{10}}} = -1.72$$

with 18 degrees of freedom and associated p value between 0.10 and 0.20. This value being large, H_0 will be retained. A 95% confidence interval on $\mu_1 - \mu_2$ is

$$(24.4 - 27.2) \pm (2.101)\sqrt{13.22225} \sqrt{\frac{1}{10} + \frac{1}{10}} = -2.8 \pm 3.4$$

Thus $-6.2 \leq \mu_1 - \mu_2 \leq 0.6$.

Inferences on Two Population Means Based on Paired Sample

In some problems, the observations taken for the first and second samples may have identical extraneous variations. For example, the data on individual's might be collected before and after some specific treatment; the pretreatment data being the first sample and the post-treatment data being the second sample. One may like to compare the IQ scores of the first and second children in a family. The family background controls the extraneous variation. The IQ scores of the first child is the first sample and the IQ scores of the second child is the second sample.

Consider a paired sample data (X_i, Y_i) for $i = 1, 2, \ldots, n$. Let μ_1 and μ_2 be the population means of the X and Y variables. Let $D_i = X_i - Y_i$. Furthermore, let $\bar{D}$ and s_d^2 be the sample mean and variance of the n differences D_is. The test statistic for testing $H_0 : \mu_1 - \mu_2 = c$ is

$$T = \frac{(\bar{D} - c)\sqrt{n}}{s_d} \tag{6A.4}$$

with $n - 1$ degrees of freedom. A $(1 - \alpha)$ 100% confidence interval on $\mu_1 - \mu_2$ is

$$\bar{D} \pm (t_{\alpha/2, n-1}) \frac{s_d}{\sqrt{n}} \tag{6A.5}$$

Example 6A.3.   Consider the data of Table 6.6. The two methods were applied on each of the five varieties and thus each pair of observations share identical extraneous variation. Thus, this is a paired sample problem. The five differences of the standard and rapid methods determination of protein content are

$$1.5, \ 1.3, \ -1.7, \ 0.4, \ -0.4$$

Here $\bar{D} = 0.22$, $s_d = 1.3142$. If μ_1 and μ_2 are the population mean protein content determined by standard and rapid methods and if the experimenter is interested in testing

$$H_0 : \ \mu_1 = \mu_2, \qquad H_A : \ \mu_1 \neq \mu_2$$

the required test statistic is

$$T = \frac{0.22\sqrt{5}}{1.3142} = 0.37$$

with 4 degrees of freedom. The p value for this statistic is more than 0.10 and H_0 will be retained. A 95% confidence interval on $\mu_1 - \mu_2$ is

$$0.22 \pm (2.776) \frac{1.3142}{\sqrt{5}} = 0.22 \pm 1.63$$

Thus $-1.41 \leq \mu_1 - \mu_2 \leq 1.85$.

Example 6A.4. Ten students in a class room were randomly selected and their scores in the first test were noted. Each of the selected students was then given an intensive coaching and their scores in the second test were recorded. The teacher wants to examine the effect of intensive coaching on test scores. This is a paired sample problem because on each student both observations were recorded. The data along with the differences are summarized in Table 6A.1. Suppose μ_1 and μ_2 are the population mean test scores without and with intensive coaching. The null and alternative hypotheses of interest to the instructor are

$$H_0: \quad \mu_1 \geq \mu_2, \qquad H_A: \quad \mu_1 < \mu_2$$

Table 6A.1   Artificial Data on Two Test Scores

Student number	First test score (X_i)	Second test score (Y_i)	$D_i = X_i - Y_i$
1	73	85	-12
2	85	90	-5
3	72	73	-1
4	58	62	-4
5	80	79	1
6	87	85	2
7	77	83	-6
8	69	75	-6
9	89	91	-2
10	73	70	3

From the data of Table 6A.1, one calculates

$$\bar{D} = -3.0, \qquad s_d = 4.55$$

The required test statistic is

$$T = \frac{(-3.0)\sqrt{10}}{4.55} = -2.09$$

with 9 degrees of freedom and p value between 0.025 and 0.05. This p-value being small, H_0 will be rejected and it will be concluded that the mean test score of the students improve with intensive coaching. A 95% confidence interval on $\mu_2 - \mu_1$ is

$$3.0 \pm 2.262 \, \frac{4.55}{\sqrt{10}} = 3.0 \pm 3.25$$

Based on this confidence interval, the improvement in average test scores will be anywhere between -0.25 and 6.25. The negative improvement can be interpreted as a decrease in test scores.

The Model

The analysis of variance is achieved through an appropriate linear model on the response variable in terms of unknown parameters explaining the causes of variation, and an error term accounting the unexplained variation.

When independent random samples are taken from k populations, the jth observation in the ith sample, X_{ij}, will be explained by

$$X_{ij} = \mu_i + e_{ij} \tag{6A.6}$$

where μ_i is the ith population mean and e_{ij} are random errors. Putting $\mu = \sum_{i=1}^{k} n_i \mu_i / n$, and $\bar{\mu}_i = \mu_i - \mu$, where n_i is the ith sample size and $n = \sum_{i=1}^{k} n_i$, X_{ij} can be rewritten as

$$X_{ij} = \mu + \bar{\mu}_i + e_{ij} \tag{6A.7}$$

One can verify that $\sum_{i=1}^{k} n_i \bar{\mu}_i = 0$. For the statistical purpose of estimation and tests of hypothesis, e_{ij} will be assumed to have independent normal distribution with mean zero and common variance σ^2.

In the case of a completely randomized design using a fixed set of k treatments, the response on the jth experimental unit receiving the ith treatment, X_{ij}, will be modeled as

$$X_{ij} = \mu + \tau_i + e_{ij} \tag{6A.8}$$

where $\sum_{i=1}^{k} n_i \tau_i = 0$, n_i being the number of EUs receiving the ith treatment. Here as all EUs are homogeneous, the common effect of these EUs for the response is denoted by μ, and τ_i represents the effect entirely due to the ith applied treatment to that EU. Even when two identical EUs are treated alike, they may show some differences in response due to unknown causes, and the random error term e_{ij} accounts for this type of variation between identically treated EUs. For the purpose of statistical analysis it will be assumed that e_{ij} are independently normally distributed with mean zero and common variance σ^2.

In a completely randomized design using k randomly selected treatments, the response X_{ij} will be modeled as

$$X_{ij} = \mu + t_i + e_{ij} \tag{6A.9}$$

where μ is the general mean, t_i is the ith treatment effect, and e_{ij} are random errors. Since the treatments are randomly selected from a population of treatments, it will be assumed that t_i are normally and independently distributed with mean 0 and variance σ_T^2. Since errors are random, e_{ij} are assumed to be normally and independently distributed with mean 0 and variance σ_E^2. It will be further assumed that t_i and e_{ij} are independent.

In a one-way analysis of variance, the number of observations under each treatment (or in each sample) need not be same. Even in a well planned experiment with equal number of replications per treatment, some observations may be lost and will not be available in course of the experiment and the experimenter will get unequal number of replications of the treatments.

In a randomized block design, using k treatments in r blocks, the response X_{ij} on the EU in the ith block receiving the jth treatment is affected by a treatment as well as a block component. When the blocks and treatments are a fixed set, X_{ij} will be modelled as

$$X_{ij} = \mu + \beta_i + \tau_j + e_{ij} \tag{6A.10}$$

where μ is a general mean, β_i is the ith block effect, τ_j is the jth treatment effect, and e_{ij} are random errors. It will be assumed that $\Sigma_{i=1}^r \beta_i = 0$, $\Sigma_{j=1}^k \tau_j = 0$. It will be further assumed that e_{ij} are normally and independently distributed with mean 0 and variance σ^2.

One-way Analysis of Variance

Using k independent samples, where the ith sample has n_i observations, the data can be summarized as follows:

		Samples		
	1	2	$\cdots$	k
	X_{11}	X_{21}	$\cdots$	X_{k1}
	X_{12}	X_{22}	$\cdots$	X_{k2}
	.	.	$\cdots$	.
	.	.	$\cdots$	.
	.	.	$\cdots$	.
	X_{1n_1}	X_{2n_2}	$\cdots$	X_{kn_k}
Totals	T_1	T_2	$\cdots$	T_k
Means	$\bar{T}_1$	$\bar{T}_2$	$\cdots$	$\bar{T}_k$

The following calculations lead to the analysis of variance table:

$$n = \sum_{i=1}^k n_i \text{ is the total number of responses}$$

$$G = \sum_{i=1}^k T_i \text{ is the total response of all } n \text{ units}$$

$$\text{Correction Factor (C.F.)} = G^2/n \qquad (6A.11)$$

Total Sum of Squares (Total SS)

$$= \sum_{i=1}^k \sum_{j=1}^{n_i} X_{ij}^2 - \text{C.F.} \ (= SS_{\text{Tot}}, \text{ say}) \qquad (6A.12)$$

Between groups SS

$$= \sum_{i=1}^{k} \frac{T_i^2}{n_i} - \text{C.F.} \ (= SS_{Bet}, \text{say}) \qquad (6A.13)$$

Within groups SS

$$= SS_{Tot} - SS_{Bet} \ (= SS_{Within}, \text{say}) \qquad (6A.14)$$

The analysis of variance table is the following:

SOURCE	DF	SS	$MS = \dfrac{SS}{DF}$	F
Between Samples	$k - 1$	SS_{Bet}	$SS_{Bet}/(k - 1)$ $= MS_{Bet}$	MS_{Bet}/MS_{Within}
Within Samples	$n - k$	SS_{Within}	$SS_{Within}/(n - k)$ $= MS_{Within}$	
Total	$n - 1$	SS_{Tot}		

The within samples mean square is, in fact, the pooled variance of the k sample variances. For a completely randomized design the calculations leading to the analysis of variance are same as before. However, the between samples component will be labeled as treatments and the within samples component will be labelled as error. The null and alternative hypotheses tested are

H_0: $\mu_1 = \mu_2 = \cdots = \mu_k$ (or $\tau_1 = \tau_2 = \cdots = \tau_k$),
H_A: at least one pair of μ_is are different (or at least one pair of τ_is are different).

The above null and alternative hypotheses can be expressed in suitable words depending on the context of the problem. The test is based on the F value calculated in the analysis of variance table. The p value for the F statistic can be determined as in the Appendix of Chapter 5, noting that this F has $k - 1$ numerator degrees of freedom and $n - k$ denominator degrees of freedom. If the p value is less than the level of significance α, it will be concluded that at least two μ_is (or τ_is) are different.

To test $\mu_i = \mu_j$ (or $\tau_i = \tau_j$) against a one-sided or a two-sided alternative, the required test statistics is

$$T = \frac{\bar{T}_i - \bar{T}_j}{\sqrt{MS_{\text{Within}}\left(\dfrac{1}{n_i} + \dfrac{1}{n_j}\right)}} \qquad (6A.15)$$

with $n - k$ degrees of freedom. The necessary p value will be calculated as described in the Appendix of Chapter 4 and conclusions drawn accordingly. The least significant difference (LSD) value for comparing the ith and jth population means (or ith and jth treatment effects) is

$$(t_{.025,n-k})\sqrt{MS_{\text{Within}}\left(\frac{1}{n_i} + \frac{1}{n_j}\right)} \qquad (6A.16)$$

In case of equal number of observations in each sample, that is, $n_1 = n_2 = \cdots = n_k$ $(= r$, say), only a single LSD value is needed and is

$$(t_{0.025,n-k})\sqrt{2MS_{\text{Within}}/r} \qquad (6A.17)$$

Two population means (or treatment effects) will be judged significant if the absolute difference of their sample means exceeds the appropriate LSD value. It may be noted that this is a two-sided test for testing the equality of the corresponding population means.

A $(1 - \alpha)$ 100% confidence interval for $\mu_i - \mu_j$ (or $\tau_j - \tau_j$) is

$$\bar{T}_i - \bar{T}_j \pm (t_{\alpha/2,n-k})\sqrt{MS_{\text{Within}}\left(\frac{1}{n_i} + \frac{1}{n_j}\right)} \qquad (6A.18)$$

In case of using a random set of treatments in a completely randomized design,

$$E(MS_{\text{Error}}) = \sigma_E^2 \qquad (6A.19)$$

$$E(MS_{\text{Treat}}) = \sigma_E^2 + \frac{1}{k-1}\left(n - \sum_{j=1}^{k}\frac{n_j^2}{n}\right)\sigma_T^2 \qquad (6A.20)$$

If $n_1 = n_2 = \cdots = n_k (= r,$ say), then (6A.20) simplifies to

$$E(MS_{Treat}) = \sigma_E^2 + r\sigma_T^2 \tag{6A.21}$$

The results discussed so far will now be illustrated through examples.

Example 6A.5. Consider Table 6.1. The four sample totals are $T_1 = 35.7$, $T_2 = 33.1$, $T_3 = 41.6$, $T_4 = 27.0$ and $G = 137.4$. Also $n_1 = 10 = n_2 = n_3 = n_4$, and $n = 40$. Now

$$CF = (137.4)^2/40 = 471.969$$

Total SS

$$\begin{aligned}
&= (3.5)^2 + (3.0)^2 + (4.5)^2 + \cdots + (1.5)^2 \\
&\quad + (3.0)^2 + (1.9)^2 - CF = 37.751
\end{aligned}$$

Between groups SS

$$= \frac{(35.7)^2}{10} + \frac{(33.1)^2}{10} + \frac{(41.6)^2}{10} + \frac{(27.0)^2}{10} - CF = 10.997$$

Within groups SS

$$= 37.751 - 10.997 = 26.754$$

Between groups MS

$$= 10.997/(4 - 1) = 3.666$$

Within groups MS

$$= 26.754/(40 - 4) = 0.743$$
$$F = 3.666/0.743 = 4.93$$

These are the values obtained on Minitab and shown in the computer output of Example 6.1. The LSD value from equation (6A.17) is

$$(1.96)\sqrt{\frac{2(0.743)}{10}} = 0.76$$

If μ_3 and μ_4 are the population mean average number of hours per day of watching television by ethnic groups C and D, a 95% confidence interval for $\mu_3 - \mu_4$ can be calculated from (6A.18) and is

$$(4.16 - 2.70) \pm (1.96)\sqrt{2(0.743)/10} = 1.46 \pm 0.76$$

Thus $0.7 \le \mu_3 - \mu_4 \le 2.22$.

   Example 6A.6.   Consider Table 6.2. Here $T_1 = 1955$, $T_2 = 1782$, $T_3 = 1640$ and $G = 5377$. Also $n_1 = 10 = n_2 = n_3$, and $n = 30$. Now

$$CF = \frac{(5377)^2}{30} = 963737.6$$

$$\text{Total SS} = (200)^2 + (195)^2 + (198)^2 + \cdots + (165)^2$$
$$+ (160)^2 + (170)^2 - CF = 5399.4$$

$$\text{Treatments SS} = \frac{(1955)^2}{10} + \frac{(1782)^2}{10} + \frac{(1640)^2}{10} - CF = 4977.3$$

$$\text{Error SS} = 5399.4 - 4977.3 = 422.1$$

$$\text{Treatments MS} = 4977.3/(3 - 1) = 2488.6$$

$$\text{Error MS} = 422.1/(30 - 3) = 15.6$$

$$F = 2488.6/15.6 = 159.19$$

These are the values obtained on Minitab and shown in the computer output of Example 6.2. The LSD value from equation (6A.17) is

$$(2.052)\sqrt{2(15.6)/10} = 3.6$$

Let τ_1 and τ_3 be the treatment effects of using 1 tablet/day and 3 tablets/day respectively. To test

$$H_0: \ \tau_3 \ge \tau_1, \qquad H_A: \ \tau_3 < \tau_1$$

the test statistic is given by (6A.15) and is

$$T = \frac{164.0 - 195.5}{\sqrt{2(15.6)}}\sqrt{10} = -17.8$$

with 27 degrees of freedom. The probability is less than 0.005 that a T

variable with 27 df is less than -17.8. A 95% confidence interval on $\tau_3 - \tau_1$ is given by (6A.18) and is

$$(164.0 - 195.5) \pm (2.052)\sqrt{2(15.6)/10} = -31.5 \pm 3.6$$

Thus $-35.1 \le \tau_3 - \tau_1 \le -27.9$.

Randomized Block Design

Consider a randomized block design in which k treatments are tested in r blocks and the data is expressed in the following tabular form

Blocks	Treatments 1	2	$\cdots$	k	Totals
1	X_{11}	X_{12}	$\cdots$	X_{1k}	B_1
2	X_{21}	X_{22}	$\cdots$	X_{2k}	B_2
.	.	.	$\cdots$	.	.
.	.	.	$\cdots$	.	.
.	.	.	$\cdots$	.	.
r	X_{r1}	X_{r2}	$\cdots$	X_{rk}	B_r
Totals	T_1	T_2	$\cdots$	T_k	G
Means	$\bar{T}_1$	$\bar{T}_2$	$\cdots$	$\bar{T}_k$	

$$CF = \frac{G^2}{kr}$$

$$\text{Total SS} = \sum_{i=1}^{r}\sum_{j=1}^{k} X_{ij}^2 - CF \ (= SS_{\text{Tot}}, \text{ say}) \tag{6A.22}$$

$$\text{Treatments SS} = \sum_{j=1}^{k} \frac{T_j^2}{r} - CF \ (= SS_{\text{Treat}}, \text{ say}) \tag{6A.23}$$

$$\text{Blocks SS} = \sum_{i=1}^{r} \frac{B_i^2}{k} - CF \ (= SS_{\text{B1}}, \text{ say}) \tag{6A.24}$$

$$\text{Error SS} = SS_{\text{Tot}} - SS_{\text{Treat}} - S_{\text{B1}} \ (= SS_{\text{Err}}, \text{ say}) \tag{6A.25}$$

The Analysis of Variance Table is then given below:

Source	DF	SS	MS = SS/DF
Treatments	$k - 1$	SS_{Treat}	$SS_{Treat}/(k - 1) = MS_{Treat}$
Blocks	$r - 1$	SS_{B1}	$SS_{B1}/(r - 1) = MS_{B1}$
Error	$(k - 1)(r - 1)$	SS_{Err}	$SS_{Err}/(k - 1)(r - 1) = MS_{Err}$
Total	$kr - 1$	SS_{Tot}	

The null and alternative hypotheses of interest to the experimenter are

H_0: $\tau_1 = \tau_2 = \cdots = \tau_k$.
H_A: at least one pair of τ_is are different.

The test statistic for this purpose is the F value given by

$$F = \frac{MS_{Treat}}{MS_{Err}} \tag{6A.26}$$

With $k - 1$ numerator degrees of freedom and $(k - 1)(r - 1)$ denominator degrees of freedom. Its p value can be easily determined as before.

The LSD value for testing any two treatment means is

$$(t_{.025,(k-1)(r-1)})\sqrt{2(MS_{Err})/r} \tag{6A.27}$$

The T statistic for testing $\tau_i = \tau_j$ for a one-sided or a two-sided test is

$$T = \frac{(\bar{T}_i - \bar{T}_j)\sqrt{r}}{\sqrt{2(MS_{Err})}} \tag{6A.28}$$

The necessary p value will be calculated as described in the Appendix of Chapter 4 and conclusions drawn in the usual manner.

A $(1 - \alpha)$ 100% confidence interval on $\tau_i - \tau_j$ is

$$(\bar{T}_i - \bar{T}_j) \pm (t_{\alpha/2,(k-1)(r-1)})\sqrt{2(MS_{Err})/r} \tag{6A.29}$$

Numerical examples illustrating the above formulae will now follow.

Example 6A.7. Consider Table 6.5. Here $r = 8$, $k = 3$, $T_1 = 245.1$, $T_2 = 229.4$, $T_3 = 240.8$, $B_1 = 85.5$, $B_2 = 89.9$, $B_3 = 91.2$, $B_4 = 93.6$, $B_5 = 87.6$, $B_6 = 86.2$, $B_7 = 92.9$, $B_8 = 88.4$, and $G = 715.3$. Now

$$CF = \frac{(715.3)^2}{24} = 21318.92$$

$$\text{Total SS} = (29.5)^2 + (29.6)^2 + (32.0)^2 + \cdots + (27.3)^2 + (30.3)^2 + (30.7)^2 - CF = 55.31$$

$$\text{Treatments SS} = \frac{(245.1)^2}{8} + \frac{(229.4)^2}{8} + \frac{(240.8)^2}{8} - CF = 16.46$$

$$\text{Blocks SS} = \frac{(85.5)^2}{3} + \frac{(89.9)^2}{3} + \frac{(91.2)^2}{3}$$

$$+ \cdots + \frac{(88.4)^2}{3} - CF = 21.02$$

$$\text{Error SS} = 55.31 - 16.46 - 21.02 = 17.83$$

$$\text{Treatments MS} = \frac{16.46}{2} = 8.23$$

$$\text{Blocks MS} = \frac{21.02}{7} = 3.00$$

$$\text{Error MS} = \frac{17.83}{14} = 1.27$$

These are the values obtained on Minitab and shown in the computer output of Example 6.7. To test

$$H_0: \quad \tau_1 = \tau_2 = \tau_3, \qquad H_A: \quad \text{Not all } \tau_i\text{s are same}$$

the test statistic is

$$F = \frac{8.23}{1.27} - 6.48$$

From equation (6A.27), the LSD value is

$$(2.145)\sqrt{2(1.27)/8} = 1.21$$

A 95% confidence interval for $\tau_1 - \tau_3$ from equation (6A.29) is

$$(30.64 - 30.1) \pm (2.145)\sqrt{2(1.27)/8} = 0.54 \pm 1.21$$

Thus $-0.67 \le \tau_1 - \tau_3 \le 1.75$.

Example 6A.8. Consider Table 6.6. Here $r = 5$, $k = 2$, $T_1 = 66.8$, $T_2 = 65.7$, $B_1 = 23.7$, $B_2 = 25.3$, $B_3 = 27.5$, $B_4 = 26.0$, $B_5 = 30.0$, and $G = 132.5$. Now

$$CF = (132.5)^2/10 = 1755.625$$

$$\text{Total SS} = (12.6)^2 + (13.3)^2 + (12.9)^2 + \cdots + (14.6)^2$$
$$+ (12.8)^2 + (15.2)^2 - CF = 14.965$$

$$\text{Treatments SS} = \frac{(66.8)^2}{5} + \frac{(65.7)^2}{5} - CF = 0.121$$

$$\text{Blocks SS} = \frac{(23.7)^2}{2} + \frac{(25.3)^2}{2} + \frac{(27.5)^2}{2} + \frac{(26.0)^2}{2}$$
$$+ \frac{(30.0)^2}{2} - CF = 11.390$$

$$\text{Error SS} = 14.965 - 0.121 - 11.390 = 3.454$$
$$\text{Treatments MS} = 0.121/1 = 0.121$$
$$\text{Blocks MS} = 11.390/4 = 2.847$$
$$\text{Error MS} = 3.454/4 = 0.863$$

These are the values obtained on Minitab and shown in the computer output of Example 6.8. To test

$$H_0: \ \tau_1 = \tau_2, \qquad H_A: \ \tau_1 \ne \tau_2$$

the required test statistic is

$$F = \frac{0.121}{0.863} = 0.14$$

A 95% confidence interval on $\tau_1 - \tau_2$ from equation (6A.29) is

$$(13.36 - 13.14) \pm (2.571)\sqrt{2(0.863)/5} = 0.22 \pm 1.51$$

Thus $-1.29 \leq \tau_1 - \tau_2 \leq 1.73$. This example was also illustrated in Example 6A.3 as a paired sample problem.

EXERCISES

1. A random sample of customers were interviewed at three departmental stores and the following data is their annual incomes (in thousands of dollars).

Stores		
A	B	C
43	18	34
18	40	29
20	35	23
42	37	28
34	16	38
32	28	34
24	20	22
37	35	15
28	22	23
25	33	20

(a) Get a Minitab output.
(b) Is there a difference in the annual incomes of customers of these 3 stores?

2. The pulse rates of a random sample of adult males and females are given below:

Males	75,	73,	76,	73,	77,	74,	76,	75,	73,	74
Females	77,	75,	76,	76,	77,	76,	76,	76,	75,	74

Perform an analysis of variance and set a 95% confidence interval for the difference of average pulse rates of adult males and females.

3. The rates of diffusion of carbon dioxide through two soils of different porosity are given below:

Fine Soil	27,	20,	29,	20,	23

Coarse Soil	30,	28,	25,	25

Get a Minitab output and test whether there is any difference between fine and coarse soils on the rates of diffusion.

4. Adaptive Behavior Scale (ABS) data on 5 patients over 4 six-monthly-periods are recorded as follows:

	ABS			
Patients	Period 1	Period 2	Period 3	Period 4
1	85	95	86	88
2	81	97	85	89
3	95	75	75	84
4	90	79	85	90
5	94	99	99	93

(a) Obtain a Minitab output.
(b) Is there any difference in average scores between the four periods?
(c) Find LSD value and find a 95% confidence interval on the average ABS scores between periods 1 and 4.
(d) Is the LSD value same as the error in estimation in a 95% confidence interval on the difference of means?

Answers

(1a)

```
ANALYSIS OF VARIANCE

SOURCE    DF    SS        MS      F
FACTOR     2      68.5    34.2    0.50
ERROR     27    1852.9    68.6
TOTAL     29    1921.4
```

(1b) No; (2) -1.2 ± 1.15; (3) not different, F value $= 1.88$; (4a)

ANALYSIS OF VARIANCE ON C1

SOURCE	DF	SS	MS
C2	3	32.4	10.8
C3	4	420.7	105.2
ERROR	12	552.1	46.0
TOTAL	19	1005.2	

(4b) No, F value $= 0.23$; (4c) LSD $= 9.35$, 0.2 ± 9.35; (4d) Yes.

7

The Use of Chi-Square Tests

Consider a frequency table using k classes for data based on a qualitative or a quantitative variable. Based on a mathematical law or prior knowledge, the investigator expects the frequencies in the k classes in the proportions $p_1, p_2, \ldots,$ and p_k. He is then interested in examining whether the deviations in observed frequencies and expected frequencies based on the expected proportions are due to chance, or real. The test statistic is a quantity called *chi square* (χ^2) based on the observed and expected frequencies and the resulting test is usually called a chi-square test. Another application of the chi-square statistic is to test the independence or dependence of the row and column classification variables in a contingency table. The equality of population proportions in two or more populations can be handled by the chi square test for a contingency table and will be considered in the text of this chapter. However, in the beginning of the Appendix, the inferences on two population proportions will be discussed and then the mathematical details of other aspects of the procedures discussed in the chapter will be given.

230

TESTS OF GOODNESS OF FIT

Let the data be the frequencies in k predetermined classes and let f_1, $f_2, \ldots, f_k$ be the observed frequencies in the k classes. By some mathematical law or based on prior knowledge, let $e_1, e_2, \ldots, e_k$ be the expected frequencies in the k classes. The classes may be grouped, if necessary, such that the expected frequencies in each class are not less than 5. The null and alternative hypotheses of interest to the investigator are

H_0: $f_1 = e_1, f_2 = e_2, \ldots, f_k = e_k$
H_A: $f_i \neq e_i$ for at least one i

or equivalently,

H_0: $p_i = p_{i0}, i = 1, 2, \ldots, k$
H_A: $p_i \neq p_{i0}$ for at least one i

where $p_1, p_2, \ldots, p_k$ are the population proportions in the k classes, and $p_{10}, p_{20}, \ldots, p_{k0}$ are the hypothesized proportions (note $\Sigma_{i=1}^{k}$ $p_i = 1$). It is sufficient to indicate the values of $k - 1$ proportions in the hypotheses. Sometimes a rephrased statement may be used to express H_0 and H_A. The test statistic here is χ^2 and its calculation and finding the p value of it are discussed in the Appendix. Conclusions on the null hypothesis are then made by the standard procedures. The following examples illustrate this test.

Example 7.1. (Unbiased Coin). To test whether a coin is unbiased, it was tossed 1000 times and heads were realized 510 times. If the coin is unbiased, one expects to get heads 500 times. While heads are expected 500 times, it is not necessary to get heads exactly 500 times due to the random nature of the outcomes in the tossing of the coin. Thus, it becomes necessary to evaluate whether getting heads 510 times is probable when heads are expected only 500 times. The null and alternative hypotheses tested in this case are

H_0: Coin is unbiased,
H_A: Coin is biased.

The outcomes as realized and expected can be shown in the following table:

	Heads	Tails	Total
Observed frequency	510	490	1000
Expected frequency	500	500	1000

The value of the test statistic χ^2 and the p value are (see Appendix)

$$\chi^2 = 0.4, \quad p\text{-value: between } 0.50 \text{ and } 0.75$$

Since the p value exceeds 0.05, it is likely to get the observed difference between the realized and expected frequencies purely by random chance. Thus H_0 is retained and it is concluded that the coin is an unbiased one.

Example 7.2.   (Classic and New Coca-Cola). Suppose, of 1200 cans of "Coke" sold in a given period in a super market, 671 are Classic and 529 are new. Does this support the hypothesis that 50% of "Coke" buyers prefer "Coke" Classic? In this problem if p denotes the proportion of Classic Coke cans sold, then H_0 and H_A are

H_0: $p = 0.5$
H_A: $p \neq 0.5$

The set-up can be represented as follows:

	Classic Coke	New Coke	Total
Observed sales	671	529	1200
Expected sales	600	600	1200

If H_0 is true, and only 600 sales of Classic Coke are expected, can the sale of 671 cans be explained by chance? The answer to this question is based on the test statistic χ^2 and its p value (see Appendix):

$$\chi^2 = 16.80, \quad p \text{ value} < .005$$

It is thus improbable that the present evidence—or stronger—can be obtained against H_0 if H_0 is true. Thus, H_0 is rejected and it is concluded that Coke drinkers do not have equal preferences for Classic and New Cokes.

Example 7.3. (Family Incomes). In a community it is believed that 10%, 20%, 30%, 30%, and 10% of families are in the income brackets <$10,000; $10,000–$20,000; $20,000–$30,000; $30,000–$40,000; and >$40,000, respectively. The observed and expected frequencies of the families in the specified income groups are tabulated:

Income group ($)	Observed frequency	Expected frequency
<10,000	6	10
10,000–20,000	31	20
20,000–30,000	34	30
30,000–40,000	24	30
>40,000	5	10
Total	100	100

To test that the observed frequencies agree with the postulated percentages, the test statistic and the p value are

$$\chi^2 = 10.32, \qquad p \text{ value: between } 0.025 \text{ and } 0.05$$

Because this p value is small, it is improbable that the present (or stronger) evidence can be obtained against the hypothesis when the hypothesis is true. Since such evidence is noted, the postulated hypothesis is rejected and it is concluded that the observed frequencies are not in agreement with the given percentages.

Example 7.4. (Mendel's Law). In an experiment on the breeding of flowers of a certain species, suppose an experimenter obtained 200 magenta flowers with a green stigma, 90 magenta flowers with a red stigma, 70 red flowers with a green stigma, and 24 red flowers with red stigma. According to Mendel's law, the theory predicts that these types should be obtained in the ratio $9:3:3:1$ or equivalently in the proportions $\frac{9}{16}$, $\frac{3}{16}$, $\frac{3}{16}$, and $\frac{1}{16}$. Thus, the observed and expected frequencies for these four types of flowers are

	Magenta-green	Magenta-red	Red-green	Red-red	Total
Observed frequency	200	90	70	24	384
Expected frequency	216	72	72	24	384

Expected frequencies are obtained by dividing the total number of flowers, 384, in the ratio $9:3:3:1$ (see Appendix). Under H_0 that the observed and expected frequencies are same except for random variations for the four types of flowers, the test statistic χ^2 and the p value are given as:

$$\chi^2 = 5.74, \qquad p \text{ value: between } 0.10 \text{ and } 0.25$$

It is thus probable that the observed frequencies for the four types of flowers can be obtained by random chance even though they are expected to occur in the ratio $9:3:3:1$. This experimental data supports Mendel's law.

CONTINGENCY TABLES

In a contingency table, the null and alternative hypotheses tested are always the following:

H_0: The variables denoting the row and column classifications are independent.

H_A: The variables are dependent.

Under H_0, the expected frequencies can be calculated as discussed in the Appendix, and the χ^2 is the necessary test statistic.

Example 7.5. (Shade and Yield). In an orchard of 1000 trees, a record was made of the number of shaded and unshaded trees, and in each of these classes the frequency of high- and low-yielding trees was noted as given in Table 7.1. This is a 2×2 contingency table because there are two rows and two columns and the frequencies are shown in the body of the table. The null and alternative hypotheses of interest to the experimenter are

H_0: The yield is independent of shading.

H_A: The yield is dependent of shading.

Table 7.1   Shade and Yield Artificial Data

	Shaded	Unshaded	Totals
Low-yielding	200	150	350
High-yielding	300	350	650
Totals	500	500	1000

The test statistic can be evaluated from the Minitab program; user's commands and computer output are shown below:

```
MTB> read c1, c2
DATA> 200 150
DATA> 300 350
DATA> end
        2 ROWS READ

MTB> chisquare c1, c2

EXPECTED FREQUENCIES ARE PRINTED BELOW OBSERVED
FREQUENCIES
          I   C1   I   C2    ITOTALS
-------I-------I-------I-------
   1  I   200  I   150  I    350
      I  175.0I  175.0I
-------I-------I-------I-------
   2  I   300  I   350  I    650
      I  325.0I  325.0I
-------I-------I-------I-------
TOTALS I   500 I   500 I   1000

TOTAL CHI SQUARE =
        3.57 + 3.57 +
        1.92 + 1.92 +
            = 10.99

DEGREES OF FREEDOM = (2-1) × (2-1) = 1
```

The χ^2 test statistic for this problem is 10.99 with 1 degree of freedom and its p value is less than 0.005. Since this p value is small, H_0 will be rejected and it is concluded that shading affects the yield.

Example 7.6. (Hair and Eye Colors). Two hundred and twenty people were classified according to their eye and hair colors and the data in Table 7.2 is obtained: This is a 3 × 3 contingency table. The null and alternative hypotheses of interest are

H_0: Eye and hair colors are independent,
H_A: Eye and hair colors are dependent.

By using Minitab as discussed in the last example, the test statistic for this problem is $\chi^2 = 16.51$ with 4 degrees of freedom and its p value is less than 0.005. Based on this, H_0 is rejected and it is concluded that eye and hair colors are dependent characters.

Table 7.2 Artificial Data on Eye and Hair Colors

	Hair color			
Eye color	Brunette	Blonde	Red	Totals
Black	40	20	5	65
Blue	20	30	10	60
Brown	60	30	5	95
Totals	120	80	20	220

Example 7.7. (Professional Competency and Education). The data in Table 7.3 were obtained from an experiment conducted to study the effect of education on the ability of hourly workers in a company. This is a 3 × 5 contingency table. The null and alternative hypotheses of interest to the experimenter are

H_0: Worker's ability is independent of the number of years of schooling.
H_A: Worker's ability is dependent of the number of years of schooling.

Table 7.3 Artificial Data on Workers' Ability and Their Education

Workers' ability	Education (years of schooling)					Totals
	8	9	10	11	12	
High	5	6	7	9	10	37
Medium	5	6	6	6	8	31
Low	5	8	9	5	5	32
Totals	15	20	22	20	23	100

As before, the reader may enter the necessary Minitab commands and get the output as follows:

```
MTB> read c1-c5
DATA> 5 6 7 9 10
DATA> 5 6 6 6 8
DATA> 5 8 9 5 5
DATA> end
       3 ROWS READ
MTB> chisquare c1-c5
```

```
EXPECTED FREQUENCIES ARE PRINTED BELOW OBSERVED
FREQUENCIES
          I   C1   I   C2   I   C3   I   C4   I   C5    ITOTALS
-------I--------I--------I--------I--------I--------I--------
   1  I   5   I   6   I   7   I   9   I  10   I    37
      I   5.6I   7.4I   8.1I   7.4I   8.5I
-------I--------I--------I--------I--------I--------I--------
   2  I   5   I   6   I   6   I   6   I   8   I    31
      I   4.7I   6.2I   6.8I   6.2I   7.1I
-------I--------I--------I--------I--------I--------I--------
   3  I   5   I   8   I   9   I   5   I   5   I    32
      I   4.8I   6.4I   7.0I   6.4I   7.4I
-------I--------I--------I--------I--------I--------I--------
TOTALS I  15  I  20  I  22  I  20  I  23  I   100
```

TOTAL CHI SQUARE =

$$.05 + .26 + .16 + .35 + .26 +$$
$$.03 + .01 + .10 + .01 + .11 +$$
$$.01 + .40 + .55 + .31 + .76 +$$

$$= 3.35$$

DEGREES OF FREEDOM = $(3-1) \times (5-1) = 8$

NOTE 2 CELLS WITH EXPECTED FREQUENCIES LESS THAN 5

From the output it will be noted that two expected frequencies are less than 5 and it is desirable to combine some classes to get a contingency table with expected frequencies that are not less than 5. This can be achieved by combining the 8 and 9 years of completed schooling into one group as "8 or 9." The table will then be a 3×4 contingency table. The necessary Minitab commands typed by the user and the output are the following:

```
MTB> let c6=c1+c2
MTB> chisquare c3-c6
```

EXPECTED FREQUENCIES ARE PRINTED BELOW OBSERVED
FREQUENCIES

	I	C3	I	C4	I	C5	I	C6	I	TOTALS
1	I	7	I	9	I	10	I	11	I	37
	I	8.1	I	7.4	I	8.5	I	13.0	I	
2	I	6	I	6	I	8	I	11	I	31
	I	6.8	I	6.2	I	7.1	I	10.9	I	.
3	I	9	I	5	I	5	I	13	I	32
	I	7.0	I	6.4	I	7.4	I	11.2	I	
TOTALS	I	22	I	20	I	23	I	35	I	100

TOTAL CHI SQUARE =

$$.16 + .35 + .26 + .29 +$$
$$.10 + .01 + .11 + .00 +$$
$$.55 + .31 + .76 + .29 +$$

$$= 3.17$$

DEGREES OF FREEDOM $= (3-1) \times (4-1) = 6$

The test statistic now is $\chi^2 = 3.17$ with 6 degrees of freedom with a p value between 0.75 and 0.90. Thus H_0 will be retained and there is no evidence from this data that workers ability depends on the years of schooling.

The type of analysis discussed in this section can be applied in a different context. Suppose there are k populations ($k \geq 2$) and suppose a sample of size n_i is taken independently from the ith population, $i = 1,2, \ldots , k$. Let each unit be classified according to its membership in a class C and let a_i be the number of units in the ith sample that belong to the class C, $i = 1,2, \ldots , k$. If p_i is the proportion of units belonging to C in the ith population, its estimate is $\hat{p}_i = a_i/n_i$, which is the sample proportion of units in C from the ith sample, $i = 1,2, \ldots , k$. The data can be conveniently summarized in a $2 \times k$ contingency table as follows:

	Sample				Totals
	1	2	$\cdots$	k	
belong to C	a_1	a_2	$\cdots$	a_k	A
does not belong to C	$n_1 - a_1$	$n_2 - a_2$	$\cdots$	$n_k - a_k$	$n - A$
Totals	n_1	n_2	$\cdots$	n_k	n

The null and alternative hypotheses of interest to the experimenter in this problem are

$$H_0: \ p_1 = p_2 = \cdots = p_k, \qquad H_A: \quad \text{at least two } p_i\text{s are different}$$

One notes that these hypotheses are equivalent to the standard hypotheses tested in a contingency table and conclusions can be drawn using the χ^2 statistic with $k - 1$ df. The following examples illustrate this analysis.

Example 7.8. (Male and Female Smokers). In a random sample of 200 males, 80 are smokers, and in a random sample of 100 females, 45 are smokers. Let p_1 and p_2 be the proportion of male and female smokers. To test

$$H_0: \quad p_1 = p_2, \qquad H_A: \quad p_1 \neq p_2$$

one summarizes the data in a 2×2 contingency table.

	Males	Females	Totals
Smokers	80	45	125
Non-smokers	120	55	175
Totals	200	100	300

The test statistic and the p-value are

$$\chi^2 = 0.69, \qquad p \text{ value between } 0.25 \text{ and } 0.50$$

Because the p value $> .05$, H_0 is retained and there is no evidence from this data set that the proportion of male smokers are different from the proportion of female smokers.

Example 7.9. (Deaths During Surgery). The number of patients operated on and the number of deaths on the operating table were noted at five hospitals in a month, giving the artificial data in Table 7.4. If p_i is the proportion of deaths on the operating table at the ith hospital, one would like to test the hypotheses $H_0: p_1 = p_2 = p_3 = p_4 = p_5$, H_A: at least a pair of p_is are different. To perform the contingency table analysis, one notes that the expected frequencies in the first row for hospitals B and E are less than five. Thus, combining adjacently labeled hospitals, a 2×3 contingency table will be considered as follows:

	Hospitals			
	A + B	C	D + E	Totals
Death	12	5	12	29
Survival	338	195	388	921
Totals	350	200	400	950

The test statistic and the p value are

$$\chi^2 = 0.38, \qquad p \text{ value between } 0.75 \text{ and } 0.90$$

Because the p value $> .05$, it is concluded that the proportion of deaths is not different in different hospitals.

Table 7.4 Deaths in Hospitals

	Hospitals					
	A	B	C	D	E	Totals
Death	8	4	5	10	2	29
Survival	192	146	195	290	98	921
Totals	200	150	200	300	100	950

APPENDIX

Inferences on Proportions for Two Populations

Suppose there are two populations. Let p_1 and p_2 be the proportions of units belonging to a specified class C in the two populations. The interest now is to test the hypotheses about p_1 and p_2 and set up a confidence interval on the difference $p_1 - p_2$. For this purpose, one may take two independent samples from these populations of large sizes n_1 and n_2 respectively. Let $\hat{p}_1$ and $\hat{p}_2$ be the two sample proportions. To test

$$H_0: \quad p_1 = p_2 \tag{7A.1}$$

one notes that under H_0, $\hat{p}_1$ and $\hat{p}_2$, each estimates the common proportion $p_1 = p_2$, and they need to be pooled together to form $\hat{p}$ given by the formula

$$\hat{p} = \frac{n_1\hat{p}_1 + n_2\hat{p}_2}{n_1 + n_2} \tag{7A.2}$$

It is easy to note that $\hat{p}$ is the proportion of sample units in C in the combined sample of size $n_1 + n_2$ from both populations. The test statistic for testing (7A.1) is the standard normal variable

$$Z = \frac{\hat{p}_1 - \hat{p}_2}{\sqrt{\hat{p}(1 - \hat{p})\left(\dfrac{1}{n_1} + \dfrac{1}{n_2}\right)}} \tag{7A.3}$$

A $(1 - \alpha)$ 100% confidence interval on $p_1 - p_2$ is given by

$$\hat{p}_1 - \hat{p}_2 \pm z_{\alpha/2} \sqrt{\frac{\hat{p}_1(1 - \hat{p}_1)}{n_1} + \frac{\hat{p}_2(1 - \hat{p}_2)}{n_2}} \tag{7A.4}$$

Instead of the null hypothesis (7A.1), if one is interested in testing

$$H_0: \quad p_1 - p_2 = c, \qquad H_A: \quad p_1 - p_2 \neq c \tag{7A.5}$$

then H_0 is rejected if c does not belong to the confidence interval given in Equation (7A.4).

Example 7A.1. Consider Example 7.8. Let p_1 and p_2 be the population proportion of smokers in males and females. The sample sizes and estimates of p_1 and p_2 are

$$n_1 = 200, \quad n_2 = 100, \quad \hat{p}_1 = \frac{80}{100} = 0.4, \quad \hat{p}_2 = \frac{45}{100} = 0.45$$

The combined estimate $\hat{p}$ is

$$\hat{p} = \frac{200(0.4) + 100(0.45)}{300} = \frac{125}{300} = 0.417$$

The test statistic for testing the hypotheses

$$H_0: \quad p_1 = p_2, \qquad H_A: \quad p_1 \neq p_2$$

is

$$Z = \frac{0.4 - 0.45}{\sqrt{(0.417)(0.583)\left(\dfrac{1}{200} + \dfrac{1}{100}\right)}} = -0.83$$

Note that this Z statistic is the square root of the χ^2 statistic of Example 7.8. The p-value for a two-sided test is

$$p\text{-value} = 2P(Z < -0.83) = 0.59$$

As this p value is large, H_0 will be retained and it will be concluded that the proportions of smokers are the same for males and females.

A 95% confidence interval on $p_1 - p_2$ is

$$0.4 - 0.45 \pm 1.96 \sqrt{\frac{(.4)(.6)}{200} + \frac{(.45)(.55)}{100}} = -0.05 \pm 0.12$$

Thus $-0.17 \leq p_1 - p_2 \leq 0.07$.

Example 7A.2. Suppose p_1 and p_2 are the proportions of male and female students in an undergraduate degree program making a grade point average of at least 3.2. A random sample of 400 male students revealed that 20 of them have a grade point average of at least 3.2. An independent random sample of 400 female students indicated that 24 of them have a grade point average of at least 3.2. Based on the data collected,

$$n_1 = 400, \qquad n_2 = 400, \qquad \hat{p}_1 = 0.05, \qquad \hat{p}_2 = 0.06$$

Since $n_1 = n_2$, $\hat{p}$ is clearly the average of $\hat{p}_1$ and $\hat{p}_2$ and is $\hat{p} = 0.055$. The test statistic for testing the hypothesis

$$H_0: \ p_1 = p_2, \qquad H_A: \ p_1 \neq p_2$$

is

$$Z = \frac{0.05 - 0.06}{\sqrt{(0.055)(0.945)\left(\dfrac{1}{400} + \dfrac{1}{400}\right)}} = -0.62$$

The p value for this test is

$$p \text{ value} = 2P(Z < -0.62) = 0.46$$

and H_0 is retained. A 95% confidence interval for $p_1 - p_2$ is

$$0.05 - 0.06 \pm (1.96)\sqrt{\frac{(0.05)(0.95)}{400} + \frac{(0.06)(0.94)}{400}}$$

$$= -0.01 \pm 0.03$$

Thus $-0.04 \leq p_1 - p_2 \leq 0.02$.

Chi-Square Distribution

The distribution of a sum of squares of independent standard normal variables is known as a chi-square distribution. This is a continuous distribution. Like the other distributions discussed in this book, this is also a very important distribution and the probabilities are tabulated in Table 4 at the end of this book. This distribution is also entered with the appropriate degrees of freedom. Rows are numbered by the degrees of freedom and tail probabilities are indicated by the columns. The chi-square values are given in the body of the table. With 1 degree of freedom, a χ^2 value of 3.84 makes a tail probability of 0.05 that is, $P(\chi^2 > 3.84) = 0.05$, because 3.84 is in the cell corresponding to row 1 and the column labeled 0.05. We write $\chi^2_{0.05,1} = 3.84$. Clearly $\chi^2_{0.01,4} = 13.3$ because 13.3 is in row 4 and column 0.01 of the table. With 2 degrees of freedom a χ^2 value of 10.0 makes a tail probability between 0.005 and 0.01 because in row 2, 10.0 occurs between the values given in columns headed by 0.01 and 0.005. A χ^2 value exceeding the last column entry makes a tail probability less than 0.005, and a χ^2 value less than the entry of the column headed by 0.995 makes a probability more than 0.995 to its right.

Test of Goodness of Fit

Consider the data arranged in a frequency table. The frequency classes may be categorical or suitable intervals of a quantitative variable. For convenience, let there be k classes denoted by $C_1, C_2, \ldots, C_k$. Let the observed frequency in these classes be respectively $f_1, f_2, \ldots, f_k$. Based on a formulated hypothesis of interest to the investigator, the frequencies in the classes are expected to be respectively $e_1, e_2, \ldots, e_k$. It is desirable to have each of the expected frequencies to be at least five, because the test statistic based on the frequency data is being approximated to a continuous distribution. In case any e_i is less than five, that frequency class can be combined with its immediately preceding or succeeding class or classes, so that the number of classes become smaller than k, and the expected frequencies in the new classification are at least five. Let $p_1, p_2, \ldots, p_k$ be the expected proportions of observations in the classes based on which e_is are calculated.

The null and alternative hypotheses of interest are

$$H_0: \quad p_i = p_{i0}, \quad i = 1, 2, \ldots, k$$
$$H_A: \quad p_i \neq p_{i0} \text{ for at least one } i$$

(7A.5)

The H_0 implies that data agrees with the postulated hypothesis. It may be noted that if the observed and expected frequencies differ in at least one class, it really means that they differ in at least two classes. It is not possible to have exactly $k - 1$ classes in which observed and expected frequencies are same.

The test statistic for testing the hypotheses of (7A.5) is

$$\chi^2 = \sum_{i=1}^{k} \frac{(f_i - e_i)^2}{e_i}$$

(7A.6)

with $k - 1$ degrees of freedom. Let c be the calculated χ^2 value from equation (7A.6). If $c < \chi^2_{0.995, k-1}$, the p value for the test is more than 0.995 and if $c > \chi^2_{0.005,k-1}$, it is less than 0.005. For c between $\chi^2_{\alpha_1,k-1}$ and $\chi^2_{\alpha_2,k-1}$ where α_1 and α_2 are two adjacent column headings, the p value is between α_1 and α_2. The following examples illustrate the calculations involved.

Example 7A.3. For the data of Example 7.1, $k = 2$, $f_1 = 510$, $e_1 = 500$, $f_2 = 490$, $e_2 = 500$. The test statistic is given by equation (7A.6) and is

$$\chi^2 = \frac{(510 - 500)^2}{500} + \frac{(490 - 500)^2}{500} = 0.4$$

with $2 - 1 = 1$ degree of freedom. In the table of the chi-square distribution, in row 1, a chi-square value of 0.4 occurs between the entries in the columns labeled 0.50 and 0.75. The p value for this test statistic based on the observed data is between 0.50 and 0.75.

Example 7A.4. For the data of Example 7.2, k is again 2 and $f_1 = 671$, $e_1 = 600$, $f_2 = 529$, $e_2 = 600$. The test statistic of equation (7A.6) is

$$\chi^2 = \frac{(671 - 600)^2}{600} + \frac{(529 - 600)^2}{600} = 16.80$$

with $2 - 1 = 1$ degree of freedom. In the chi-square table, in row 1, a chi-square value of 16.80 exceeds the entry in the column headed by 0.005 and thus the required p value is less than 0.005.

Example 7A.5. Consider the data of Example 7.4. The total of 384 plants must belong to the four plant types in a ratio of $9:3:3:1$. Thus the proportion of plants of the four types are respectively $\frac{9}{16}$, $\frac{3}{16}$, $\frac{3}{16}$, and $\frac{1}{16}$, because out of $9 + 3 + 3 + 1 = 16$ plants, 9, 3, 3, and 1 plants belong to the four types. The expected number of plants in the four types are respectively $(384)(\frac{9}{16}) = 216$, $(384)(\frac{3}{16}) = 72$, $(384)(\frac{3}{16}) = 72$, and $(384)(\frac{1}{16}) = 24$. These expected frequencies are given in that example. Here $k = 4$, $f_1 = 200$, $e_1 = 216$, $f_2 = 90$, $e_2 = 72$, $f_3 = 70$, $e_3 = 72$, $f_4 = 24$, and $e_4 = 24$. The test statistic of equation (7A.6) is

$$\chi^2 = \frac{(200 - 216)^2}{216} + \frac{(90 - 72)^2}{72} + \frac{(70 - 72)^2}{72} + \frac{(24 - 24)^2}{24}$$

$$= 5.74$$

with $4 - 1 = 3$ degrees of freedom. In row 3 of Table 4 at the end of this book, a chi-square value of 5.74 occurs between the entries in columns headed by 0.10 and 0.25. The p value in this example is between 0.10 and 0.25.

Contingency Table

A contingency table is a row-column frequency table where the frequencies are indicated based on two variables. If the table has r rows and c columns, it is known as a $r \times c$ contingency table. For clarity, let A denote the row classification and B denote the column classification. The observed frequencies can then be shown as follows:

	Factor B				
Factor A	1	2	$\cdots$	c	Totals
1	f_{11}	f_{12}	$\cdots$	f_{1c}	A_1
2	f_{21}	f_{22}	$\cdots$	f_{2c}	A_2
.	.	.	$\cdots$	.	.
.	.	.	$\cdots$	.	.
.	.	.	$\cdots$	.	.
r	f_{r1}	f_{r2}	$\cdots$	f_{rc}	A_r
Totals	B_1	B_2	$\cdots$	B_c	n

The null and alternative hypotheses tested here are always

$$H_0: \quad \text{Factors A and B are independent}$$
$$H_A: \quad \text{Factors A and B are dependent} \tag{7A.7}$$

An equivalent statement with the same interpretation as H_0 and H_A can also be used for H_0 and H_A. Under H_0, if e_{ij} is the expected frequency in the cell (i,j), it follows that the proportion of observations in the jth column must be same in each row, that is,

$$\frac{e_{1j}}{A_1} = \frac{e_{2j}}{A_2} = \frac{e_{3j}}{A_3} = \cdots = \frac{e_{rj}}{A_r} \tag{7A.8}$$

Equation (7A.8) implies that

$$\frac{e_{ij}}{A_i} = \frac{B_j}{n} \qquad (7A.9)$$

and hence

$$e_{ij} = \frac{(A_i)(B_j)}{n} \qquad (7A.10)$$

for all i and j. Once the expected frequencies are obtained, H_0 and H_A of (7A.7) are equivalent to

$$H_0: \ f_{ij} = e_{ij}, \text{ for all } i,j$$
$$H_A: \ f_{ij} \neq e_{ij} \text{ for at least one cell } (i,j) \qquad (7A.11)$$

The necessary test statistic is again

$$\chi^2 = \sum_{i=1}^{r} \sum_{j=1}^{c} \frac{(f_{ij} - e_{ij})^2}{e_{ij}} \qquad (7A.12)$$

with $(r - 1)(c - 1)$ degrees of freedom. The p value can then be determined and inferences drawn in the usual manner. These ideas will now be illustrated.

Example 7A.6. Consider the 2×2 contingency table discussed in Example 7.5. Here

$$f_{11} = 200, \qquad f_{12} = 150, \qquad A_1 = 350$$
$$f_{21} = 300, \qquad f_{22} = 350, \qquad A_2 = 650$$
$$B_1 = 500, \qquad B_2 = 500, \qquad n = 1000$$

From equation (7A.10), the expected frequencies are given by

$$e_{11} = \frac{(350)(500)}{1000} = 175, \qquad e_{12} = \frac{(350)(500)}{1000} = 175$$

$$e_{21} = \frac{(650)(500)}{1000} = 325, \qquad e_{22} = \frac{(650)(500)}{1000} = 325$$

The calculated test statistic from equation (7A.12) is

$$\chi^2 = \frac{(200 - 175)^2}{175} + \frac{(150 - 175)^2}{175} + \frac{(300 - 325)^2}{325}$$

$$+ \frac{(350 - 325)^2}{325} = 10.99$$

with $(2 - 1)(2 - 1) = 1$ degree of freedom. In row 1 of the chi-square tables, a chi-square value of 10.99 exceeds the entry given in the column headed 0.005 and thus the p value in this example is less than 0.005.

Example 7A.7. For the 3 × 3 contingency table discussed in Example 7.6,

$$
\begin{array}{llll}
f_{11} = 40, & f_{12} = 20, & f_{13} = 5, & A_1 = 65 \\
f_{21} = 20, & f_{22} = 30, & f_{23} = 10, & A_2 = 60 \\
f_{31} = 60, & f_{32} = 30, & f_{33} = 5, & A_3 = 95 \\
B_1 = 120, & B_2 = 80, & B_3 = 20, & n = 220
\end{array}
$$

From equation (7A.10), the expected frequencies are given by

$$e_{11} = (65)(120)/220 = 35.5, \qquad e_{12} = (65)(80)/220 = 23.6,$$
$$e_{13} = (65)(20)/220 = 5.9$$
$$e_{21} = (60)(120)/220 = 32.7, \qquad e_{22} = (60)(80)/220 = 21.8,$$
$$e_{23} = (60)(20)/220 = 5.5$$
$$e_{31} = (95)(120)/220 = 51.8, \qquad e_{32} = (95)(80)/220 = 34.5,$$
$$e_{33} = (95)(20)/220 = 8.6$$

The evaluated test statistic from equation (7A.12) is

$$\chi^2 = \frac{(40 - 35.5)^2}{35.5} + \frac{(20 - 23.6)^2}{23.6} + \frac{(5 - 5.9)^2}{5.9} + \frac{(20 - 32.7)^2}{32.7}$$

$$+ \frac{(30 - 21.8)^2}{21.8} + \frac{(10 - 5.5)^2}{5.5} + \frac{(60 - 51.8)^2}{51.8}$$

$$+ \frac{(30 - 34.5)^2}{34.5} + \frac{(5 - 8.6)^2}{8.6} = 16.51$$

with $(3 - 1)(3 - 1) = 4$ degrees of freedom. In row 4 of the table, a chi-square value of 16.51 exceeds the entry in the column 0.005 and hence the appropriate p value here is less than 0.005.

For a 2×2 contingency table

			Totals
	a	b	$a + b$
	c	d	$c + d$
Totals	$a + c$	$b + d$	$a + b + c + d = n$

χ^2 with 1 degree of freedom is given by the formula

$$\chi^2 = \frac{(ad - bc)^2 n}{(a + b)(c + d)(a + c)(b + d)} \qquad (7A.13)$$

EXERCISES

1. The number of books borrowed from a public library during week-days are

Day	Mon	Tues	Wed	Thurs	Fri
No. books borrowed	240	160	200	150	250

Is there evidence that the number of books borrowed are different on different days?

2. Certain cross-breedings of peas gave 460 yellow and 140 green seeds. According to Mendel's law these colors should occur in a $3 : 1$ ratio. Does the data support Mendel's theory?

3. In a random sample of 400 women aged 40 and over, five have breast cancer. Is there evidence that 2% of women of that age group have breast cancer?

4. The following artificial data is related to the segregation of two genes for purple-red flower color and long-round pollen shape in sweet peas:

	Color		
Shape	Purple	Red	Totals
Long	296	27	323
Round	19	8	27
Totals	315	35	350

Is there evidence that color and shape are dependent characters?

5. The following artificial data indicate the number of premarital sexual experiences in different races

	Premarital Sex	
Races	Yes	No
A	50	50
B	40	60
C	15	35
D	10	40

Is there evidence that the same rate of premarital sex is prevalent in different races?

6. The following artificial data indicate the income status of parents and the highest education obtained by any of their children:

Parents' income	Highest education of any child				
	Not completed high school	Completed high school	Bachelor's	Master's	Doctorate
Low	40	160	60	35	5
Middle	20	100	180	70	30
High	5	55	100	30	10

Is the highest education of children dependent on parental income level?

7. In a classroom of 40 students there are 24 males and 16 females. 10 males and 12 females received a B grade or better. Is there evidence to conclude that the proportions of B grade (or better) students differ in males and females?

Answers

(1) Yes, $\chi^2 = 41.00$, p value < 0.005; (2) Yes, $\chi^2 = 0.89$, p value: between 0.25 and 0.50; (3) Yes, $\chi^2 = 1.15$, p value: between 0.25 and 0.50; (4) Yes, $\chi^2 = 12.53$, p value < 0.005; (5) No, $\chi^2 = 14.45$, p value: between 0.005 and 0.01; (6) Yes, $\chi^2 = 121.96$, p value < 0.005; (7) Yes, $\chi^2 = 4.31$, p value: between 0.025 and 0.05.

Review Exercises

1. The net profits made by an used car dealer on five cars are
   $100, $20, $50, $250, $100
   The mean profit per car made by the dealer is
   (a) $104   (b) $100   (c) $250
   (d) none of these
2. The calculated s for the net profits for the data of Exercise 1 is
   (a) $50   (b) $88.49   (c) $200
   (d) none of these
3. The median profit for the data of Exercise 1 is
   (a) $50   (b) $200   (c) $100   (d) none of these
4. The number of defective items found in several boxes have a mean of 2.5 and a mode of 1. The curve approximating the histogram of the number of defective items per box is
   (a) symmetrical   (b) right skewed   (c) left skewed
5. The gas mileage given by brand X cars is normally distributed with a mean of 20 mpg and a S.D. of 2 mpg. Is it probable for a brand X car to deliver a gas mileage of 27 mpg?
   (a) Yes   (b) No

6. The test scores given by an instructor are normally distributed
    with a mean of 65. The highest and lowest test scores are 87.5
    and 42.5. What percentage of students received scores between
    72.5 and 87.5?
    (a) 2½%   (b) 13½%   (c) 16%
    (d) none of these
7. The weights of 20 to 25 year old adult females follow a
    (a) normal distribution   (b) right skewed distribution
    (c) left skewed distribution
8. Can $\bar{X} = s^2$ for any data set?
    (a) Yes   (b) No
9. The median family income in a community is 20,000 dollars. Is
    it true that there are more people making more than 20,000
    dollars than people making less than that amount in that commu-
    nity?
    (a) Always true   (b) Possibly true   (c) Not true
10. The probability that a major league baseball team will win and
    not win the next year's World Series is
    (a) 0.5   (b) 1   (c) 0   (d) none of these
11. In a three-children family, the probability that all children are of
    the same sex is
    (a) ⅛   (b) ⅞   (c) ⅝   (d) none of these
12. The percentage of people who will buy a newly developed prod-
    uct is theoretically a
    (a) continuous variable   (b) discrete variable
13. When a pair of dice are thrown, the probability of a total of 6 is
    (a) ⁶⁄₃₆   (b) ¹⁄₃₆   (c) ⁵⁄₃₆   (d) none of these
14. If A and B are two independent events such that $P(A) = 0.2$,
    $P(B) = 0.3$, then $P(A$ or $B)$ is
    (a) 0.06   (b) 0.44   (c) 0.55   (d) none of these
15. Is it more probable to get a head when you toss a single coin, or
    an even number when you throw a dice?
    (a) Head is more probable   (b) Tail is more probable
    (c) Both are the same
    Use the following setting to answer Exercises 16–18.
    The profit made by a publisher in publishing a book, X, has the
    following pdf:

X	−$10,000	−$5,000	0	$100,000
p(x)	0.8	0.05	0.05	0.1

16. The expected profit made by the publisher on a new book is
    (a) $1750   (b) −$1750   (c) $100,000
    (d) none of these
17. Do you recommend that the publisher publish such books?
    (a) Yes   (b) No
18. The probability of losing money by the publisher in such a venture is
    (a) 0.8   (b) 0.85   (c) 0.95   (d) none of these
19. Let X represent the number of times a housewife visits a local grocery store in a one-week period. The pdf of X is given as:

X	0	1	2	3
p(X)	0.1	0.3	0.4	0.2

What is her expected number of visits to the grocery store?
    (a) 1.7   (b) 3   (c) 2   (d) none of these
20. The probability of getting a 6 five times when you throw a dice 10 times is

(a) $\binom{10}{5}\left(\frac{1}{6}\right)^{10}$ (b) $\binom{10}{5}\left(\frac{1}{6}\right)^{5}\left(\frac{5}{6}\right)^{5}$ (c) $\binom{10}{6}\left(\frac{1}{5}\right)^{6}$

    (d) none of these
21. If the probability that a person likes Diet Slice is 0.3, does it mean that 30% of the population likes Diet Slice?
    (a) No   (b) Yes
22. If three people flip coins, an odd man is determined if one flips a head while the other two flip a tail, or if one flips a tail while the other two flip a head. What is the probability that an odd man is not determined in a game when all three people throw the coins only once?
    (a) ¼   (b) ¾   (c) ½   (d) none of these

23. If students in a class receive an average score of 70 points with a standard deviation of 8 points, the student receiving 58 is
    (a) 1.5 SD away from the mean
    (b) 1.5 SD above the mean
    (c) 1.5 SD below the mean
    (d) none of these

24. The gas mileage of brand X cars is normally distributed with a mean of 21.0 mpg, and a standard deviation of 0.5 mpg. Bob has a brand X car that gives 22.0 mpg. What proportion of brand X cars give a gas mileage equal to or better than Bob's car?
    (a) 2.5%   (b) 2.25%   (c) 0.0227%
    (d) none of these

25. Bob received a test score of 85 points when his class average score was 80 with a standard deviation of 2 points. John received a test score of 86 points when his class average was 80, with a standard deviation of 3 points. Which of the two is a better student?
    (a) John   (b) Bob

26. The test scores in a class are normally distributed with a mean of 75 points and a standard deviation of 7 points. What should be the bottom score required for an A grade if only 5% of the students are given an A grade? (Round off to integers)
    (a) 89   (b) 87   (c) 82   (d) none of these

27. A 90% confidence interval for μ is 10 ± 1.5. The investigator took a census of the entire population and found μ to actually equal 12.0. Are the confidence interval calculations wrong?
    (a) No   (b) Yes   (c) Not necessarily

28. A random sample of 1000 homes found that 999 have television sets. A 95% confidence interval for the proportion of all homes having television sets is
    (a) 0.999 ± 0.002 (b) 0.999 ± 0.062
    (c) 0.999 ± 1.96 (d) none of these

29. Is it true that we must double the sample size if we want to halve the size of the error in estimation?
    (a) Yes   (b) No

30. A random sample of 100 students in a college have an average age $\bar{X} = 20$ years with a standard deviation $s = 5$ years. A 80% confidence interval for the average age of all students in that

college is
(a) 20 ± 0.98   (b) 20 ± 0.64   (c) 20 ± 5
(d) none of these

31. The weights (in pounds) of five randomly selected teenage boys are

128   140   125   148   145

A 90% confidence interval for the weight of all teenage boys is
(a) 137.2 ± 7.53   (b) 137.2 ± 9.75
(c) 137.2 ± 10.23   (d) none of these

32. A consumer research group believes that the lives of television picture tubes are normally distributed with a standard deviation of 3 years. How many sets must they test if they want to be 95% certain that their estimate of the mean life will be within 1.5 years?
(a) 25   (b) 16   (c) 1844   (d) None of these

33. How many people should be examined to estimate the proportion of people with AIDS to within 3 percentage points using a 99% confidence interval? (No guess value of this proportion is available.)
(a) 100   (b) 184   (c) 1844   (d) None of these

34. A p value of 0.0501 was calculated for testing H_0: $\mu = 12{,}000$, H_A: $\mu \neq 12{,}000$. Do we retain H_0 using a 0.05 level of significance?
(a) Yes   (b) No

35. Two independent agencies conducted polls regarding the usefulness of the 1985 US and Soviet summit meeting and tested the hypothesis H_0: $p \leq 0.5$, H_A: $p > 0.5$, where p is the proportion of people who consider the meeting useful. The first agency calculated a p value of 0.0207, and the second agency calculated a p value of 0.0192. Which agency poll indicates a stronger evidence for rejecting H_0?
(a) First   (b) Second

36. The stopping distance for a company's premium tire is 54 m with a standard deviation of 10 m. Twenty-five of these tires were tested and the average stopping distance of the tested tires was 48 m. Is there evidence that these tires have stopping distances less than 54 m (give p value with your answer)?
(a) Yes, p value =   (b) No, p value =

37. Is it possible to reject a H_0 in favor of a one-sided alternative, when H_0 is retained against a two-sided alternative?
    (a) Yes   (b) No
38. A firm operating a large number of vending machines studies the relation between maintenance cost and dollar sales for each of their machines. The purpose is to identify those machines the cost of which were "out of line" with their sales volume. The dependent variable for the regression equation is
    (a) Maintenance cost   (b) Sales volume
39. A multiple regression equation $\hat{Y} = b_0 + b_1 X + b_2 X^2$ was fitted with an $R^2 = 0.99$. Is the correlation between X and Y, $\sqrt{0.99}$?
    (a) Yes   (b) No
40. A multiple regression equation $\hat{Y} = 250 - 700X_1 + 100X_2 + 5X_3 + 15X_1 X_3$ was fitted. If X_2 and X_3 are held constant, is it true that $\hat{Y}$ decreases by 700 on an average for each unit increase in X_1?
    (a) Yes   (b) No
41. Is it necessary that the independent variable with largest absolute value for the regression coefficient be the most important variable in a regression equation?
    (a) Yes   (b) No

Use the following partial Minitab output obtained by regressing the gas mileage of a car (Y) for the average speed during a trip (X) for seven identically equipped same model cars, to answer questions 42–46.

THE REGRESSION EQUATION IS

$$\hat{Y} = 72.3 - 0.960 \times$$

COLUMN	COEFFICIENT	ST. DEV. OF COEF.	T-RATIO = COEF/S.D.
	72.32	17.78	4.07
X	−0.9602	0.3049	−3.15

ANALYSIS OF VARIANCE

DUE TO	DF	SS	MS = SS/DF
REGRESSION	1	120.51	120.51
RESIDUAL	5	60.75	12.15
TOTAL	6	181.26	

42. The percentage of variation in Y explained by the regression equation is

    (a) 66.5   (b) 33.5   (c) 9.9   (d) none of these

43. The estimated s for this data is

    (a) 7.79   (b) 3.49   (c) 13.46

    (d) none of these

44. Is there statistical evidence that cars give better gas mileage at lower speeds?

    (a) Yes, p value $=$ (b) No, p value $=$

45. The expected gas mileage delivered by such cars at 55 mph is

    (a) 19.5   (b) 52.8   (c) 17.78

    (d) none of these

46. The fitted equation is a

    (a) poor predictor   (b) good predictor

    (c) excellent predictor

47. The assumptions used in one-way Anova are

    (a) samples are independent

    (b) populations are normally distributed

    (c) populations have some variances

    (d) all the three above

    (e) none of these

48. The time taken (in seconds) to finish a calculation was tested on three calculators using three students giving the following data:

Student #	Calculator Brand		
	A	B	C
1	60	50	70
2	70	60	80
3	50	60	55

To test the effectiveness of the three brands of calculators, we use

    (a) CRD analysis   (b) RBD analysis

    (c) contingency table analysis   (d) none of these

49. An educational psychologist is studying differences in intelligence scores produced by two test procedures A and B. Five randomly selected subjects were each given both tests A and B with the following results:

Subject	Test Score in	
	A	B
1	81	93
2	103	70
3	79	88
4	85	91
5	84	83

The differences between the test procedures can be analyzed by the analysis of a

(a) contingency table   (b) CRD   (c) RBD

(d) none of these

50. An investment analyst is studying the relation between stock price movements in 2 weeks for 100 stocks and obtained the following data

Movement in 1st week	Movement in 2nd week		Totals
	Increase	No Increase	
Increase	21	19	40
No Increase	29	31	60
Totals	50	50	100

Are the first- and second-week movement of the stocks independent?

(a) Yes, p value =   (b) No, p value =

51. In a random sample of 100 males in the age group of 40 to 50 years, 15 had had a heart attack. In a random sample 100 females in the same age group, eight had had a heart attack. Is there statistical evidence to conclude that the same proportion of males and females have heart attacks in that age group?

(a) Yes, p value =   (b) No, p value =

52. In a health clinic, a diabetes test was conducted on random samples of 100 whites, 100 blacks, 50 Hispanics and 20 Asians. Ten whites, 15 blacks, 8 Hispanics and 4 Asians were positive for the test. Is there statistical evidence to conclude that the proportion of diabetes cases are same in the four groups?
   (a) Yes, p value =   (b) No, p value =

53. The number of sales on each of the 5 days in the week in a store are as follows:

Weekday	Mon.	Tues.	Wed.	Thurs.	Fri.
No. sales	50	20	40	30	60

   Is there statistical evidence to conclude that the number of sales are different on the five weekdays?
   (a) Yes, p value =   (b) No, p value =

54. There are 51,000 men and 49,000 women in a community. If that community is considered as a sample of human population, is there statistical evidence to conclude that men and women are in equal proportions?
   (a) Yes, p value =   (b) No, p value =

55. χ^2 value in a contingency table is zero if
   (a) row and column classifications are independent
   (b) row and column classifications are dependent
   (c) none of these

56. What is the appropriate statistical analysis to test that gestation depends on the average life of animals for the data given in Table 3.7?
   (a) Correlation analysis   (b) RBD analysis
   (c) Contingency table analysis   (d) None of these

57. Combine the last two rows and last two columns in the main body of Table 3.7. Is there statistical evidence to conclude that gestation depends on the average life of animals?
   (a) Yes, p value =   (h) No, p value =

58. Combine the columns corresponding to D, F, and W grades in Table 3.9 to make it a 3 × 4 contingency table. Is there evidence that the grades given are independent of the teachers who taught

the course?

(a)   Yes, p value =   (b)   No, p value =

59.   Perform RBD analysis on the data of Table 3.14 and test whether the temperatures have statistically significant effects on the ascorbic acid content.

(a)   Significant   (b)   Not significant

60.   Change the percentages in Table 3.16 to the actual and expected frequencies in the classes and test the goodness of fit of the actual frequencies with the expected frequencies.

(a)   Fit is significant, p value =

(b)   Fit is not significant, p value =

Appendix

Table A.1   Stanard Normal Distribution Function

$$N(t) = \int_{-\infty}^{t} \frac{1}{\sqrt{2\pi}} e^{-z^2/2} \, dz$$

t	0	1	2	3	4	5	6	7	8	9
−3.	.0013	.0013	.0013	.0012	.0012	.0011	.0011	.0011	.0010	.0010
−2.9	.0019	.0018	.0017	.0017	.0016	.0016	.0015	.0015	.0014	.0014
−2.8	.0026	.0025	.0024	.0023	.0023	.0022	.0021	.0021	.0020	.0019
−2.7	.0035	.0034	.0033	.0032	.0031	.0030	.0029	.0028	.0027	.0026
−2.6	.0047	.0045	.0044	.0043	.0041	.0040	.0039	.0038	.0037	.0036
−2.5	.0062	.0060	.0059	.0057	.0055	.0054	.0052	.0051	.0049	.0048
−2.4	.0082	.0080	.0078	.0075	.0073	.0071	.0069	.0068	.0066	.0064
−2.3	.0107	.0104	.0102	.0099	.0096	.0094	.0091	.0089	.0087	.0084
−2.2	.0139	.0136	.0132	.0129	.0125	.0122	.0119	.0116	.0113	.0110
−2.1	.0179	.0174	.0170	.0166	.0162	.0158	.0154	.0150	.0146	.0143
−2.0	.0227	.0222	.0217	.0212	.0207	.0202	.0197	.0192	.0188	.0183
−1.9	.0287	.0281	.0274	.0268	.0262	.0256	.0250	.0244	.0239	.0233
−1.8	.0359	.0351	.0344	.0336	.0329	.0322	.0314	.0307	.0300	.0294
−1.7	.0446	.0436	.0427	.0418	.0409	.0401	.0392	.0384	.0375	.0367
−1.6	.0548	.0537	.0526	.0516	.0505	.0495	.0485	.0475	.0465	.0455
−1.5	.0668	.0655	.0643	.0630	.0618	.0606	.0594	.0582	.0571	.0559
−1.4	.0808	.0793	.0778	.0764	.0749	.0735	.0721	.0708	.0694	.0681
−1.3	.0968	.0951	.0934	.0918	.0901	.0885	.0869	.0853	.0838	.0823
−1.2	.1151	.1131	.1112	.1093	.1075	.1056·	.1038	.1020	.1003	.0985
−1.1	.1357	.1335	.1314	.1292	.1271	.1251	.1230	.1210	.1190	.1170
−1.0	.1587	.1562	.1539	.1515	.1492	.1469	.1446	.1423	.1401	.1379
−.9	.1841	.1814	.1788	.1762	.1736	.1711	.1685	.1660	.1635	.1611
−.8	.2119	.2090	.2061	.2033	.2005	.1977	.1949	.1921	.1894	.1867
−.7	.2420	.2389	.2358	.2326	.2297	.2266	.2236	.2206	.2177	.2148
−.6	.2743	.2709	.2676	.2643	.2611	.2578	.2546	.2514	.2483	.2451
−.5	.3085	.3050	.3015	.2981	.2946	.2912	.2877	.2843	.2810	.2776
−.4	.3446	.3409	.3372	.3336	.3300	.3264	.3228	.3192	.3156	.3121
−.3	.3821	.3783	.3745	.3707	.3669	.3632	.3594	.3557	.3520	.3483
−.2	.4207	.4168	.4129	.4090	.4052	.4013	.3974	.3936	.3897	.3859
−.1	.4602	.4562	.4522	.4483	.4443	.4404	.4364	.4325	.4286	.4247
−.0	.5000	.4960	.4920	.4880	.4840	.4801	.4761	.4721	.4681	.4641

Table A.1 Continued

t	0	1	2	3	4	5	6	7	8	9
.0	.5000	.5040	.5080	.5120	.5160	.5199	.5239	.5279	.5319	.5359
.1	.5398	.5438	.5478	.5517	.5557	.5596	.5636	.5675	.5714	.5753
.2	.5793	.5832	.5871	.5910	.5948	.5987	.6026	.6064	.6103	.6141
.3	.6179	.6217	.6255	.6293	.6331	.6368	.6406	.6443	.6480	.6517
.4	.6554	.6591	.6628	.6664	.6700	.6736	.6772	.6808	.6844	.6879
.5	.6915	.6950	.6985	.7019	.7054	.7088	.7123	.7157	.7190	.7224
.6	.7257	.7291	.7324	.7357	.7389	.7422	.7454	.7486	.7517	.7549
.7	.7580	.7611	.7642	.7673	.7704	.7734	.7764	.7794	.7823	.7852
.8	.7881	.7910	.7939	.7967	.7995	.8023	.8051	.8079	.8106	.8133
.9	.8159	.8186	.8212	.8238	.8264	.8289	.8315	.8340	.8365	.8389
1.0	.8413	.8438	.8461	.8485	.8508	.8531	.8554	.8577	.8599	.8621
1.1	.8643	.8665	.8686	.8708	.8729	.8749	.8770	.8790	.8810	.8830
1.2	.8849	.8869	.8888	.8907	.8925	.8944	.8962	.8980	.8997	.9015
1.3	.9032	.9049	.9066	.9082	.9099	.9115	.9131	.9147	.9162	.9177
1.4	.9192	.9207	.9222	.9236	.9251	.9265	.9279	.9292	.9306	.9319
1.5	.9332	.9345	.9357	.9370	.9382	.9394	.9406	.9418	.9429	.9441
1.6	.9452	.9463	.9474	.9484	.9495	.9505	.9515	.9525	.9535	.9545
1.7	.9554	.9564	.9573	.9582	.9591	.9599	.9608	.9616	.9625	.9633
1.8	.9641	.9649	.9656	.9664	.9671	.9678	.9686	.9693	.9700	.9706
1.9	.9713	.9719	.9726	.9732	.9738	.9744	.9750	.9756	.9761	.9767
2.0	.9773	.9778	.9783	.9788	.9793	.9798	.9803	.9808	.9812	.9817
2.1	.9821	.9826	.9830	.9834	.9838	.9842	.9846	.9850	.9854	.9857
2.2	.9861	.9864	.9868	.9871	.9875	.9878	.9881	.9884	.9887	.9890
2.3	.9893	.9896	.9898	.9901	.9904	.9906	.9909	.9911	.9913	.9916
2.4	.9918	.9920	.9922	.9925	.9927	.9929	.9931	.9932	.9934	.9936
2.5	.9938	.9940	.9941	.9943	.9945	.9946	.9948	.9949	.9951	.9952
2.6	.9953	.9955	.9956	.9957	.9959	.9960	.9961	.9962	.9963	.9964
2.7	.9965	.9966	.9967	.9968	.9969	.9970	.9971	.9972	.9973	.9974
2.8	.9974	.9975	.9976	.9977	.9977	.9978	.9979	.9979	.9980	.9981
2.9	.9981	.9982	.9982	.9983	.9984	.9984	.9985	.9985	.9986	.9986
3.	.9987	.9987	.9987	.9988	.9988	.9989	.9989	.9989	.9990	.9990

Source: Reproduced from Larsen, *Introduction to Probability Theory and Statistical Inference,* pp. 580–581, with permission of the author, and publisher John Wiley & Sons, Inc.

Table A.2   Values of Student's *t*

d.f.	$t_{.100}$	$t_{.050}$	$t_{.025}$	$t_{.010}$	$t_{.005}$
1	3.078	6.314	12.706	31.821	63.657
2	1.886	2.920	4.303	6.965	9.925
3	1.638	2.353	3.182	4.541	5.841
4	1.533	2.132	2.776	3.747	4.604
5	1.476	2.015	2.571	3.365	4.032
6	1.440	1.943	2.447	3.143	3.707
7	1.415	1.895	2.365	2.998	3.499
8	1.397	1.860	2.306	2.896	3.355
9	1.383	1.833	2.262	2.821	3.250
10	1.372	1.812	2.228	2.764	3.169
11	1.363	1.796	2.201	2.718	3.106
12	1.356	1.782	2.179	2.681	3.055
13	1.350	1.771	2.160	2.650	3.012
14	1.345	1.761	2.145	2.624	2.977
15	1.341	1.753	2.131	2.602	2.947
16	1.337	1.746	2.120	2.583	2.921
17	1.333	1.740	2.110	2.567	2.898
18	1.330	1.734	2.101	2.552	2.878
19	1.328	1.729	2.093	2.539	2.861
20	1.325	1.725	2.086	2.528	2.845
21	1.323	1.721	2.080	2.518	2.831
22	1.321	1.717	2.074	2.508	2.819
23	1.319	1.714	2.069	2.500	2.807
24	1.318	1.711	2.064	2.492	2.797
25	1.316	1.708	2.060	2.485	2.787
26	1.315	1.706	2.056	2.479	2.779
27	1.314	1.703	2.052	2.473	2.771
28	1.313	1.701	2.048	2.467	2.763
29	1.311	1.699	2.045	2.462	2.756
inf.	1.282	1.645	1.960	2.326	2.576

Source: Reproduced from Table of percentage points of the *t*-distribution, *Biometrika* 32:300 (1941), by M. Merrington with permission of the editor of *Biometrika*.

Table A.3   5% (Roman type) and 1% (Boldface type) Points for the F Distribution

ν_1, df in Numerator

ν_2	1	2	3	4	5	6	7	8	9	10	11	12	14	16	20	24	30	40	50	75	100	200	500	∞
1	161 **4,052**	200 **4,999**	216 **5,403**	225 **5,625**	230 **5,764**	234 **5,859**	237 **5,928**	239 **5,981**	241 **6,022**	242 **6,056**	243 **6,082**	244 **6,106**	245 **6,142**	246 **6,169**	248 **6,208**	249 **6,234**	250 **6,261**	251 **6,286**	252 **6,302**	253 **6,323**	253 **6,334**	254 **6,352**	254 **6,361**	254 **6,366**
2	18.51 **98.49**	19.00 **99.00**	19.16 **99.17**	19.25 **99.25**	19.30 **99.30**	19.33 **99.33**	19.36 **99.36**	19.37 **99.37**	19.38 **99.39**	19.39 **99.40**	19.40 **99.41**	19.41 **99.42**	19.42 **99.43**	19.43 **99.44**	19.44 **99.45**	19.45 **99.46**	19.46 **99.47**	19.47 **99.48**	19.47 **99.48**	19.48 **99.49**	19.49 **99.49**	19.49 **99.49**	19.50 **99.50**	19.50 **99.50**
3	10.13 **34.12**	9.55 **30.82**	9.28 **29.46**	9.12 **28.71**	9.01 **28.24**	8.94 **27.91**	8.88 **27.67**	8.84 **27.49**	8.81 **27.34**	8.78 **27.23**	8.76 **27.13**	8.74 **27.05**	8.71 **26.92**	8.69 **26.83**	8.66 **26.69**	8.64 **26.60**	8.62 **26.50**	8.60 **26.41**	8.58 **26.35**	8.57 **26.27**	8.56 **26.23**	8.54 **26.18**	8.54 **26.14**	8.53 **26.12**
4	7.71 **21.20**	6.94 **18.00**	6.59 **16.69**	6.39 **15.98**	6.26 **15.52**	6.16 **15.21**	6.09 **14.98**	6.04 **14.80**	6.00 **14.66**	5.96 **14.54**	5.93 **14.45**	5.91 **14.37**	5.87 **14.24**	5.84 **14.15**	5.80 **14.02**	5.77 **13.93**	5.74 **13.83**	5.71 **13.74**	5.70 **13.69**	5.68 **13.61**	5.66 **13.57**	5.65 **13.52**	5.64 **13.48**	5.63 **13.46**
5	6.61 **16.26**	5.79 **13.27**	5.41 **12.06**	5.19 **11.39**	5.05 **10.97**	4.95 **10.67**	4.88 **10.45**	4.82 **10.29**	4.78 **10.15**	4.74 **10.05**	4.70 **9.96**	4.68 **9.89**	4.64 **9.77**	4.60 **9.68**	4.56 **9.55**	4.53 **9.47**	4.50 **9.38**	4.46 **9.29**	4.44 **9.24**	4.42 **9.17**	4.40 **9.13**	4.38 **9.07**	4.37 **9.04**	4.36 **9.02**
6	5.99 **13.74**	5.14 **10.92**	4.76 **9.78**	4.53 **9.15**	4.39 **8.75**	4.28 **8.47**	4.21 **8.26**	4.15 **8.10**	4.10 **7.98**	4.06 **7.87**	4.03 **7.79**	4.00 **7.72**	3.96 **7.60**	3.92 **7.52**	3.87 **7.39**	3.84 **7.31**	3.81 **7.23**	3.77 **7.14**	3.75 **7.09**	3.72 **7.02**	3.71 **6.99**	3.69 **6.94**	3.68 **6.90**	3.67 **6.88**
7	5.59 **12.25**	4.74 **9.55**	4.35 **8.45**	4.12 **7.85**	3.97 **7.46**	3.87 **7.19**	3.79 **7.00**	3.73 **6.84**	3.68 **6.71**	3.63 **6.62**	3.60 **6.54**	3.57 **6.47**	3.52 **6.35**	3.49 **6.27**	3.44 **6.15**	3.41 **6.07**	3.38 **5.98**	3.34 **5.90**	3.32 **5.85**	3.29 **5.78**	3.28 **5.75**	3.25 **5.70**	3.24 **5.67**	3.23 **5.65**
8	5.32 **11.26**	4.46 **8.65**	4.07 **7.59**	3.84 **7.01**	3.69 **6.63**	3.58 **6.37**	3.50 **6.19**	3.44 **6.03**	3.39 **5.91**	3.34 **5.82**	3.31 **5.74**	3.28 **5.67**	3.23 **5.56**	3.20 **5.48**	3.15 **5.36**	3.12 **5.28**	3.08 **5.20**	3.05 **5.11**	3.03 **5.06**	3.00 **5.00**	2.98 **4.96**	2.96 **4.91**	2.94 **4.88**	2.93 **4.86**
9	5.12 **10.56**	4.26 **8.02**	3.86 **6.99**	3.63 **6.42**	3.48 **6.06**	3.37 **5.80**	3.29 **5.62**	3.23 **5.47**	3.18 **5.35**	3.13 **5.26**	3.10 **5.18**	3.07 **5.11**	3.02 **5.00**	2.98 **4.92**	2.93 **4.80**	2.90 **4.73**	2.86 **4.64**	2.82 **4.56**	2.80 **4.51**	2.77 **4.45**	2.76 **4.41**	2.73 **4.36**	2.72 **4.33**	2.71 **4.31**
10	4.96 **10.04**	4.10 **7.56**	3.71 **6.55**	3.48 **5.99**	3.33 **5.64**	3.22 **5.39**	3.14 **5.21**	3.07 **5.06**	3.02 **4.95**	2.97 **4.85**	2.94 **4.78**	2.91 **4.71**	2.86 **4.60**	2.82 **4.52**	2.77 **4.41**	2.74 **4.33**	2.70 **4.25**	2.67 **4.17**	2.64 **4.12**	2.61 **4.05**	2.59 **4.01**	2.56 **3.96**	2.55 **3.93**	2.54 **3.91**
11	4.84 **9.65**	3.98 **7.20**	3.59 **6.22**	3.36 **5.67**	3.20 **5.32**	3.09 **5.07**	3.01 **4.88**	2.95 **4.74**	2.90 **4.63**	2.86 **4.54**	2.82 **4.46**	2.79 **4.40**	2.74 **4.29**	2.70 **4.21**	2.65 **4.10**	2.61 **4.02**	2.57 **3.94**	2.53 **3.86**	2.50 **3.80**	2.47 **3.74**	2.45 **3.70**	2.42 **3.66**	2.41 **3.62**	2.40 **3.60**
12	4.75 **9.33**	3.88 **6.93**	3.49 **5.95**	3.26 **5.41**	3.11 **5.06**	3.00 **4.82**	2.92 **4.65**	2.85 **4.50**	2.80 **4.39**	2.76 **4.30**	2.72 **4.22**	2.69 **4.16**	2.64 **4.05**	2.60 **3.98**	2.54 **3.86**	2.50 **3.78**	2.46 **3.70**	2.42 **3.61**	2.40 **3.56**	2.36 **3.49**	2.35 **3.46**	2.32 **3.41**	2.31 **3.38**	2.30 **3.36**
13	4.67 **9.07**	3.80 **6.70**	3.41 **5.74**	3.18 **5.20**	3.02 **4.86**	2.92 **4.62**	2.84 **4.44**	2.77 **4.30**	2.72 **4.19**	2.67 **4.10**	2.63 **4.02**	2.60 **3.96**	2.55 **3.85**	2.51 **3.78**	2.46 **3.67**	2.42 **3.59**	2.38 **3.51**	2.34 **3.42**	2.32 **3.37**	2.28 **3.30**	2.26 **3.27**	2.24 **3.21**	2.22 **3.18**	2.21 **3.16**

Table A.3 Continued

v_1, df in Numerator

v_2	1	2	3	4	5	6	7	8	9	10	11	12	14	16	20	24	30	40	50	75	100	200	500	∞
14	4.60/8.86	3.74/6.51	3.34/5.56	3.11/5.03	2.96/4.69	2.85/4.46	2.77/4.28	2.70/4.14	2.65/4.03	2.60/3.94	2.56/3.86	2.53/3.80	2.48/3.70	2.44/3.62	2.39/3.51	2.35/3.43	2.31/3.34	2.27/3.26	2.24/3.21	2.21/3.14	2.19/3.11	2.16/3.06	2.14/3.02	2.13/3.00
15	4.54/8.68	3.68/6.36	3.29/5.42	3.06/4.89	2.90/4.56	2.79/4.32	2.70/4.14	2.64/4.00	2.59/3.89	2.55/3.80	2.51/3.73	2.48/3.67	2.43/3.56	2.39/3.48	2.33/3.36	2.29/3.29	2.25/3.20	2.21/3.12	2.18/3.07	2.15/3.00	2.12/2.97	2.10/2.92	2.08/2.89	2.07/2.87
16	4.49/8.53	3.63/6.23	3.24/5.29	3.01/4.77	2.85/4.44	2.74/4.20	2.66/4.03	2.59/3.89	2.54/3.78	2.49/3.69	2.45/3.61	2.42/3.55	2.37/3.45	2.33/3.37	2.28/3.25	2.24/3.18	2.20/3.10	2.16/3.01	2.13/2.96	2.09/2.96	2.07/2.86	2.04/2.80	2.02/2.77	2.01/2.75
17	4.45/8.40	3.59/6.11	3.20/5.18	2.96/4.67	2.81/4.34	2.70/4.10	2.62/3.93	2.55/3.79	2.50/3.68	2.45/3.59	2.41/3.52	2.38/3.45	2.33/3.35	2.29/3.27	2.23/3.16	2.19/3.08	2.15/3.00	2.11/2.92	2.08/2.86	2.04/2.79	2.02/2.76	1.99/2.70	1.97/2.67	1.96/2.65
18	4.41/8.28	3.55/6.01	3.16/5.09	2.93/4.58	2.77/4.25	2.66/4.01	2.58/3.85	2.51/3.71	2.46/3.60	2.41/3.51	2.37/3.44	2.34/3.37	2.29/3.27	2.25/3.19	2.19/3.07	2.15/3.00	2.11/2.91	2.07/2.83	2.04/2.78	2.00/2.71	1.98/2.68	1.95/2.62	1.93/2.59	1.92/2.57
19	4.38/8.18	3.52/5.93	3.13/5.01	2.90/4.50	2.74/4.17	2.63/3.94	2.55/3.77	2.48/3.63	2.43/3.52	2.38/3.43	2.34/3.36	2.31/3.30	2.26/3.19	2.21/3.12	2.15/3.00	2.11/2.92	2.07/2.84	2.02/2.76	2.00/2.70	1.96/2.63	1.94/2.60	1.91/2.54	1.90/2.51	1.88/2.49
20	4.35/8.10	3.49/5.85	3.10/4.94	2.87/4.43	2.71/4.10	2.60/3.87	2.52/3.71	2.45/3.56	2.40/3.45	2.35/3.37	2.31/3.30	2.28/3.23	2.23/3.13	2.18/3.05	2.12/2.94	2.08/2.86	2.04/2.77	1.99/2.69	1.96/2.63	1.92/2.56	1.90/2.53	1.87/2.47	1.85/2.44	1.84/2.42
21	4.32/8.02	3.47/5.78	3.07/4.87	2.84/4.37	2.68/4.04	2.57/3.81	2.49/3.65	2.42/3.51	2.37/3.40	2.32/3.31	2.28/3.24	2.25/3.17	2.20/3.07	2.15/2.99	2.09/2.88	2.05/2.80	2.00/2.72	1.96/2.63	1.93/2.58	1.89/2.51	1.87/2.47	1.84/2.42	1.82/2.38	1.81/2.36
22	4.30/7.94	3.44/5.72	3.05/4.82	2.82/4.31	2.66/3.99	2.55/3.76	2.47/3.59	2.40/3.45	2.35/3.35	2.30/3.26	2.26/3.18	2.23/3.12	2.18/3.02	2.13/2.94	2.07/2.83	2.03/2.75	1.98/2.67	1.93/2.58	1.91/2.53	1.87/2.46	1.84/2.42	1.81/2.37	1.80/2.33	1.78/2.31
23	4.28/7.88	3.42/5.66	3.03/4.76	2.80/4.26	2.64/3.94	2.53/3.71	2.45/3.54	2.38/3.41	2.32/3.30	2.28/3.21	2.24/3.14	2.20/3.07	2.14/2.97	2.10/2.89	2.04/2.78	2.00/2.70	1.96/2.62	1.91/2.53	1.88/2.48	1.84/2.41	1.82/2.37	1.79/2.32	1.77/2.28	1.76/2.26
24	4.26/7.82	3.40/5.61	3.01/4.72	2.78/4.22	2.62/3.90	2.51/3.67	2.43/3.50	2.36/3.36	2.30/3.25	2.26/3.17	2.22/3.09	2.18/3.03	2.13/2.93	2.09/2.85	2.02/2.74	1.98/2.66	1.94/2.58	1.89/2.49	1.86/2.44	1.82/2.36	1.80/2.33	1.76/2.27	1.74/2.23	1.73/2.21
25	4.24/7.77	3.38/5.57	2.99/4.68	2.76/4.18	2.60/3.86	2.49/3.63	2.41/3.46	2.34/3.32	2.28/3.21	2.24/3.13	2.20/3.05	2.16/2.99	2.11/2.89	2.06/2.81	2.00/2.70	1.96/2.62	1.92/2.54	1.87/2.45	1.84/2.40	1.80/2.32	1.77/2.29	1.74/2.23	1.72/2.19	1.71/2.17
26	4.22/7.72	3.37/5.53	2.98/4.64	2.74/4.14	2.59/3.82	2.47/3.59	2.39/3.42	2.32/3.29	2.27/3.17	2.22/3.09	2.18/3.02	2.15/2.96	2.10/2.86	2.05/2.77	1.99/2.66	1.95/2.58	1.90/2.50	1.85/2.41	1.82/2.36	1.78/2.28	1.76/2.25	1.72/2.19	1.70/2.15	1.69/2.13

Top column labels (left to right): 27 28 29 30 32 34 36 38 40 42 44 46 48

df†																								
27	1.67/2.10	1.68/2.12	1.71/2.16	1.74/2.21	1.76/2.25	1.80/2.33	1.84/2.38	1.88/2.47	1.93/2.55	1.97/2.63	2.03/2.74	2.08/2.83	2.13/2.93	2.16/2.98	2.20/3.06	2.25/3.14	2.30/3.26	2.37/3.39	2.46/3.56	2.57/3.79	2.73/4.11	2.96/4.60	3.35/5.49	4.21/7.68
28	1.65/2.06	1.67/2.09	1.69/2.13	1.72/2.18	1.75/2.22	1.78/2.30	1.81/2.35	1.87/2.44	1.91/2.52	1.96/2.60	2.02/2.71	2.06/2.80	2.12/2.90	2.15/2.95	2.19/3.03	2.24/3.11	2.29/3.23	2.36/3.36	2.44/3.53	2.56/3.76	2.71/4.07	2.95/4.57	3.34/5.45	4.20/7.64
29	1.64/2.03	1.65/2.06	1.68/2.10	1.71/2.15	1.73/2.19	1.77/2.27	1.80/2.32	1.85/2.41	1.90/2.49	1.94/2.57	2.00/2.68	2.05/2.77	2.10/2.87	2.14/2.92	2.18/3.00	2.22/3.08	2.28/3.20	2.35/3.33	2.43/3.50	2.54/3.73	2.70/4.04	2.93/4.54	3.33/5.42	4.18/7.60
30	1.62/2.01	1.64/2.03	1.66/2.07	1.69/2.13	1.72/2.16	1.76/2.24	1.79/2.29	1.84/2.38	1.89/2.47	1.93/2.55	1.99/2.66	2.04/2.74	2.09/2.84	2.12/2.90	2.16/2.98	2.21/3.06	2.27/3.17	2.34/3.30	2.42/3.47	2.53/3.70	2.69/4.02	2.92/4.51	3.32/5.39	4.17/7.56
32	1.59/1.96	1.61/1.98	1.64/2.02	1.67/2.08	1.69/2.12	1.74/2.20	1.76/2.25	1.82/2.34	1.86/2.42	1.91/2.51	1.97/2.62	2.02/2.70	2.07/2.80	2.10/2.86	2.14/2.94	2.19/3.01	2.25/3.12	2.32/3.25	2.40/3.42	2.51/3.66	2.67/3.97	2.90/4.46	3.30/5.34	4.15/7.50
34	1.57/1.91	1.59/1.94	1.61/1.98	1.64/2.04	1.67/2.08	1.71/2.15	1.74/2.21	1.80/2.30	1.84/2.38	1.89/2.47	1.95/2.58	2.00/2.66	2.05/2.76	2.08/2.82	2.12/2.89	2.17/2.97	2.23/3.08	2.30/3.21	2.38/3.38	2.49/3.61	2.65/3.93	2.88/4.42	3.28/5.29	4.13/7.44
36	1.55/1.87	1.56/1.90	1.59/1.94	1.62/2.00	1.65/2.04	1.69/2.12	1.72/2.17	1.78/2.26	1.82/2.35	1.87/2.43	1.93/2.54	1.98/2.62	2.03/2.72	2.06/2.78	2.10/2.86	2.15/2.94	2.21/3.04	2.28/3.18	2.36/3.35	2.48/3.58	2.63/3.89	2.86/4.38	3.26/5.25	4.11/7.39
38	1.53/1.84	1.54/1.86	1.57/1.90	1.60/1.97	1.63/2.00	1.67/2.08	1.71/2.14	1.76/2.22	1.80/2.32	1.85/2.40	1.92/2.51	1.96/2.59	2.02/2.69	2.05/2.75	2.09/2.82	2.14/2.91	2.19/3.02	2.26/3.15	2.35/3.32	2.46/3.54	2.62/3.86	2.85/4.34	3.25/5.21	4.10/7.35
40	1.51/1.81	1.53/1.84	1.55/1.88	1.59/1.94	1.61/1.97	1.66/2.05	1.69/2.11	1.74/2.20	1.79/2.29	1.84/2.37	1.90/2.49	1.95/2.56	2.00/2.66	2.04/2.73	2.07/2.80	2.12/2.88	2.18/2.99	2.25/3.12	2.34/3.29	2.45/3.51	2.61/3.83	2.84/4.31	3.23/5.18	4.08/7.31
42	1.49/1.78	1.51/1.80	1.54/1.85	1.57/1.91	1.60/1.94	1.64/2.02	1.68/2.08	1.73/2.17	1.78/2.26	1.82/2.35	1.89/2.46	1.94/2.54	1.99/2.64	2.02/2.70	2.06/2.77	2.11/2.86	2.17/2.96	2.24/3.10	2.32/3.26	2.44/3.49	2.59/3.80	2.83/4.29	3.22/5.15	4.07/7.27
44	1.48/1.75	1.50/1.78	1.52/1.82	1.56/1.88	1.58/1.92	1.63/2.00	1.66/2.06	1.72/2.15	1.76/2.24	1.81/2.32	1.88/2.44	1.92/2.52	1.98/2.62	2.01/2.68	2.05/2.75	2.10/2.84	2.16/2.94	2.23/3.07	2.31/3.24	2.43/3.46	2.58/3.78	2.82/4.26	3.21/5.12	4.06/7.24
46	1.46/1.72	1.48/1.76	1.51/1.80	1.54/1.86	1.57/1.90	1.62/1.98	1.65/2.04	1.71/2.13	1.75/2.22	1.80/2.30	1.87/2.42	1.91/2.50	1.97/2.60	2.00/2.66	2.04/2.73	2.09/2.82	2.15/2.92	2.22/3.05	2.30/3.22	2.42/3.44	2.57/3.76	2.81/4.24	3.20/5.10	4.05/7.21
48	1.45/1.70	1.47/1.73	1.50/1.78	1.53/1.84	1.56/1.88	1.61/1.96	1.64/2.02	1.70/2.11	1.74/2.20	1.79/2.28	1.86/2.40	1.90/2.48	1.96/2.58	1.99/2.64	2.03/2.71	2.08/2.80	2.14/2.90	2.21/3.04	2.30/3.20	2.41/3.42	2.56/3.74	2.80/4.22	3.19/5.08	4.04/7.19

†df in denominator.

Table A.3 Continued

n_2†	\ n_1, df in Numerator	1	2	3	4	5	6	7	8	9	10	11	12	14	16	20	24	30	40	50	75	100	200	500	∞
50		4.03/7.17	3.18/5.06	2.79/4.20	2.56/3.72	2.40/3.41	2.29/3.18	2.20/3.02	2.13/2.88	2.07/2.78	2.02/2.70	1.98/2.62	1.95/2.56	1.90/2.46	1.85/2.39	1.78/2.26	1.74/2.18	1.69/2.10	1.63/2.00	1.60/1.94	1.55/1.86	1.52/1.82	1.48/1.76	1.46/1.71	1.44/1.68
55		4.02/7.12	3.17/5.01	2.78/4.16	2.54/3.68	2.38/3.37	2.27/3.15	2.18/2.98	2.11/2.85	2.05/2.75	2.00/2.66	1.97/2.59	1.93/2.53	1.88/2.43	1.83/2.35	1.76/2.23	1.72/2.15	1.67/2.06	1.61/1.96	1.58/1.90	1.52/1.82	1.50/1.78	1.46/1.71	1.43/1.66	1.41/1.64
60		4.00/7.08	3.15/4.98	2.76/4.13	2.52/3.65	2.37/3.34	2.25/3.12	2.17/2.95	2.10/2.82	2.04/2.72	1.99/2.63	1.95/2.56	1.92/2.50	1.86/2.40	1.81/2.32	1.75/2.20	1.70/2.12	1.65/2.03	1.59/1.93	1.56/1.87	1.50/1.79	1.48/1.74	1.44/1.68	1.41/1.63	1.39/1.60
65		3.99/7.04	3.14/4.95	2.75/4.10	2.51/3.62	2.36/3.31	2.24/3.09	2.15/2.93	2.08/2.79	2.02/2.70	1.98/2.61	1.94/2.54	1.90/2.47	1.85/2.37	1.80/2.30	1.73/2.18	1.68/2.09	1.63/2.00	1.57/1.90	1.54/1.84	1.49/1.76	1.46/1.71	1.42/1.64	1.39/1.60	1.37/1.56
70		3.98/7.01	3.13/4.92	2.74/4.08	2.50/3.60	2.35/3.29	2.23/3.07	2.14/2.91	2.07/2.77	2.01/2.67	1.97/2.59	1.93/2.51	1.89/2.45	1.84/2.35	1.79/2.28	1.72/2.15	1.67/2.07	1.62/1.98	1.56/1.88	1.53/1.82	1.47/1.74	1.45/1.69	1.40/1.62	1.37/1.56	1.35/1.53
80		3.96/6.96	3.11/4.88	2.72/4.04	2.48/3.56	2.33/3.25	2.21/3.04	2.12/2.87	2.05/2.74	1.99/2.64	1.95/2.55	1.91/2.48	1.88/2.41	1.82/2.32	1.77/2.24	1.70/2.11	1.65/2.03	1.60/1.94	1.54/1.84	1.51/1.78	1.45/1.70	1.42/1.65	1.38/1.57	1.35/1.52	1.32/1.49
100		3.94/6.90	3.09/4.82	2.70/3.98	2.46/3.51	2.30/3.20	2.19/2.99	2.10/2.82	2.03/2.69	1.97/2.59	1.92/2.51	1.88/2.43	1.85/2.36	1.79/2.26	1.75/2.19	1.68/2.06	1.63/1.98	1.57/1.89	1.51/1.79	1.48/1.73	1.42/1.64	1.39/1.59	1.34/1.51	1.30/1.46	1.28/1.43
125		3.92/6.84	3.07/4.78	2.68/3.94	2.44/3.47	2.29/3.17	2.17/2.95	2.08/2.79	2.01/2.65	1.95/2.56	1.90/2.47	1.86/2.40	1.83/2.33	1.77/2.23	1.72/2.15	1.65/2.03	1.60/1.94	1.55/1.85	1.49/1.75	1.45/1.68	1.39/1.59	1.36/1.54	1.31/1.46	1.27/1.40	1.25/1.37
150		3.91/6.81	3.06/4.75	2.67/3.91	2.43/3.44	2.27/3.14	2.16/2.92	2.07/2.76	2.00/2.62	1.94/2.53	1.89/2.44	1.85/2.37	1.82/2.30	1.76/2.20	1.71/2.12	1.64/2.00	1.59/1.91	1.54/1.83	1.47/1.72	1.44/1.66	1.37/1.56	1.34/1.51	1.29/1.43	1.25/1.37	1.22/1.33
200		3.89/6.76	3.04/4.71	2.65/3.88	2.41/3.41	2.26/3.11	2.14/2.90	2.05/2.73	1.98/2.60	1.92/2.50	1.87/2.41	1.83/2.34	1.80/2.28	1.74/2.17	1.69/2.09	1.62/1.97	1.57/1.88	1.52/1.79	1.45/1.69	1.42/1.62	1.35/1.53	1.32/1.48	1.26/1.39	1.22/1.33	1.19/1.28
400		3.86/6.70	3.02/4.66	2.62/3.83	2.39/3.36	2.23/3.06	2.12/2.85	2.03/2.69	1.96/2.55	1.90/2.46	1.85/2.37	1.81/2.29	1.78/2.23	1.72/2.12	1.67/2.04	1.60/1.92	1.54/1.84	1.49/1.74	1.42/1.64	1.38/1.57	1.32/1.47	1.28/1.42	1.22/1.32	1.16/1.24	1.13/1.19
1000		3.85/6.66	3.00/4.62	2.61/3.80	2.38/3.34	2.22/3.04	2.10/2.82	2.02/2.66	1.95/2.53	1.89/2.43	1.84/2.34	1.80/2.26	1.76/2.20	1.69/2.09	1.65/2.01	1.58/1.89	1.53/1.81	1.46/1.71	1.41/1.61	1.36/1.54	1.30/1.44	1.26/1.38	1.19/1.28	1.13/1.19	1.08/1.11
∞		3.84/6.63	2.99/4.60	2.60/3.78	2.37/3.32	2.21/3.02	2.09/2.80	2.01/2.64	1.94/2.51	1.88/2.41	1.83/2.32	1.79/2.24	1.75/2.18	1.69/2.07	1.64/1.99	1.57/1.87	1.52/1.79	1.46/1.69	1.40/1.59	1.35/1.52	1.28/1.41	1.24/1.36	1.17/1.25	1.11/1.15	1.00/1.00

† df in denominator.

Source: Reproduced from Snedecor and Cochran, Statistical Methods, 6th Ed., pp. 560–563, with the kind permission of the Iowa State University Press, Ames, Iowa.

Table A.4 Values of χ^2

						Probability of a larger value of χ^2							
df	.995	.990	.975	.950	.900	.750	.500	.250	.100	.050	.025	.010	.005
1	$.0^4393$	$.0^3157$	$.0^3982$	$.0^3393$	.0158	.102	.455	1.32	2.71	3.84	5.02	6.63	7.88
2	.0100	.0201	.0506	.103	.211	.575	1.39	2.77	4.61	5.99	7.38	9.21	10.6
3	.0717	.115	.216	.352	.584	1.21	2.37	4.11	6.25	7.81	9.35	11.3	12.8
4	.207	.297	.484	.711	1.06	1.92	3.36	5.39	7.78	9.49	11.1	13.3	14.9
5	.412	.554	.831	1.15	1.61	2.67	4.35	6.63	9.24	11.1	12.8	15.1	16.7
6	.676	.872	1.24	1.64	2.20	3.45	5.35	7.84	10.6	12.6	14.4	16.8	18.5
7	.989	1.24	1.69	2.17	2.83	4.25	6.35	9.04	12.0	14.1	16.0	18.5	20.3
8	1.34	1.65	2.18	2.73	3.49	5.07	7.34	10.2	13.4	15.5	17.5	20.1	22.0
9	1.73	2.09	2.70	3.33	4.17	5.90	8.34	11.4	14.7	16.9	19.0	21.7	23.6
10	2.16	2.56	3.25	3.94	4.87	6.74	9.34	12.5	16.0	18.3	20.5	23.2	25.2
11	2.60	3.05	3.82	4.57	5.58	7.58	10.3	13.7	17.3	19.7	21.9	24.7	26.8
12	3.07	3.57	4.40	5.23	6.30	8.44	11.3	14.8	18.5	21.0	23.3	26.2	28.3
13	3.57	4.11	5.01	5.89	7.04	9.30	12.3	16.0	19.8	22.4	24.7	27.7	29.8
14	4.07	4.66	5.63	6.57	7.79	10.2	13.3	17.1	21.1	23.7	26.1	29.1	31.3
15	4.60	5.23	6.26	7.26	8.55	11.0	14.3	18.2	22.3	25.0	27.5	30.6	32.8
16	5.14	5.81	6.91	7.96	9.31	11.9	15.3	19.4	23.5	26.3	28.8	32.0	34.3
17	5.70	6.41	7.56	8.67	10.1	12.8	16.3	20.5	24.8	27.6	30.2	33.4	35.7
18	6.26	7.01	8.23	9.39	10.9	13.7	17.3	21.6	26.0	28.9	31.5	34.8	37.2
19	6.84	7.63	8.91	10.1	11.7	14.6	18.3	22.7	27.2	30.1	32.9	36.2	38.6
20	7.43	8.26	9.59	10.9	12.4	15.5	19.3	23.8	28.4	31.4	34.2	37.6	40.0
21	8.03	8.90	10.3	11.6	13.2	16.3	20.3	24.9	29.6	32.7	35.5	38.9	41.4
22	8.64	9.54	11.0	12.3	14.0	17.2	21.3	26.0	30.8	33.9	36.8	40.3	42.8
23	9.26	10.2	11.7	13.1	14.8	18.1	22.3	27.1	32.0	35.2	38.1	41.6	44.2
24	9.89	10.9	12.4	13.8	15.7	19.0	23.3	28.2	33.2	36.4	39.4	43.0	45.6
25	10.5	11.5	13.1	14.6	16.5	19.9	24.3	29.3	34.4	37.7	40.6	44.3	46.9
26	11.2	12.2	13.8	15.4	17.3	20.8	25.3	30.4	35.6	38.9	41.9	45.6	48.3
27	11.8	12.9	14.6	16.2	18.1	21.7	26.3	31.5	36.7	40.1	43.2	47.0	49.6
28	12.5	13.6	15.3	16.9	18.9	22.7	27.3	32.6	37.9	41.3	44.5	48.3	51.0
29	13.1	14.3	16.0	17.7	19.8	23.6	28.3	33.7	39.1	42.6	45.7	49.6	52.3
30	13.8	15.0	16.8	18.5	20.6	24.5	29.3	34.8	40.3	43.8	47.0	50.9	53.7
40	20.7	22.2	24.4	26.5	29.1	33.7	39.3	45.6	51.8	55.8	59.3	63.7	66.8
50	28.0	29.7	32.4	34.8	37.7	42.9	49.3	56.3	63.2	67.5	71.4	76.2	79.5
60	35.5	37.5	40.5	43.2	46.5	52.3	59.3	67.0	74.4	79.1	83.3	88.4	92.0

Source: reproduced from Steel and Torrie, *Principles and Procedures of Statistics*, p. 435, abridged from C. M. Thompson (1941), Table of percentage points of the χ^2 distribution. *Biometrica* 32:188–189, with permission of the editor of *Biometrica*.

Index